THERMODYNAMICS

Y V C Rao

Universities Press

Universities Press (India) Private Limited

Registered Office
3-6-747/1/A & 3-6-754/1, Himayatnagar, Hyderabad 500 029 (A.P), India
e-mail: *www.universitiespress.com*

Distributed by
Orient Blackswan Private Limited

Registered Office
3-6-752, Himayatnagar, Hyderabad 500 029 (A.P), India

Other Offices
Bangalore / Bhopal / Bhubaneshwar / Chennai /Ernakulam / Guwahati
Hyderabad / Jaipur / Kolkata /Lucknow / Mumbai / New Delhi / Patna

First published 2001
Reprint 2013

ISBN: 978 81 7371 388 0

Typeset by
OSDATA, Hyderabad 500 029

Printed at
Graphica Printers and Binders
Hyderabad 500 013

Published by
Universities Press (India) Private Limited
3-6-747/1/A & 3-6-754/1, Himayatnagar,
Hyderabad 500 029 (A.P), India

Dedicated to
Lord Sri Venkateswara

Contents

PREFACE

Thermodynamics deals with equilibrium, energy, and its transformation from one form to another. The laws of thermodynamics are of wide applicability and are used in several branches of engineering and science. Thermodynamics is one of the most fascinating and important branches of engineering science and is imparted as a compulsory core course (a course which is common to all branches of engineering) to undergraduate engineering students in almost all universities and engineering colleges.

It is well known that different universities and engineering colleges follow a slightly different syllabus for thermodynamics. Within the same syllabus even, emphasis on certain topics may change from university to university. In many engineering colleges, elementary thermodynamics is taught at the first year level and the subject is usually assigned 22–25 lecture hours. After teaching thermodynamics for the past 28 years at all levels—core, professional and postgraduate—at IIT Kanpur, I felt the need to prepare a textbook on thermodynamics to cater to the requirements of different universities and engineering colleges.

Keeping the time limitation (the number of lecture hours assigned) in mind, this book has been prepared to present the essential concepts and principles without any loss of rigor. Particular emphasis has been placed on precise and logical presentation of the concepts and principles which are essential to the understanding of thermodynamics. The text contains numerous solved examples to help students in understanding the underlying concepts and principles. In addition, there are several review questions and problems with answers at the end of each chapter to enable the students to gain confidence in applying the principles of thermodynamics.

Almost all the chapter end problems can be solved by direct application of the principles presented in the text. However, for the benefit of students and teachers a solutions manual containing complete worked-out solutions to all the chapter end problems is being prepared. I hope that this text will contribute to effective teaching of thermodynamics at the core level and that students will derive the maximum benefit from it. Some errors might have inadvertently crept into this book and I would greatly appreciate if the readers would bring them to my notice.

I would like to gratefully acknowledge the financial support provided by the Curriculum Development Cell of the Quality Improvement Programme, IIT Kanpur, of the Ministry of Human Resource Development, for the preparation of the manuscript. I should like to express my appreciation to my wife Mrs Y V Ramana who faced the arduous task of typing the rough draft, in a very short time, from my hand written notes. Thanks are due also to my children Srinivas and Sawjanya who have been a source of great encouragement to me in completing this task. Finally, I wish to sincerely thank the publishers, Universities Press (India) Ltd., for publishing this book in a very short time.

Kanpur
June 2001

Y V C Rao

NOMENCLATURE

a van der Waals constant
a Acceleration
b van der Waals constant
COP Coefficient of performance
C_p Specific heat at constant pressure
C_v Specific heat at constant volume
E Energy
e Specific energy
F Force
H Enthalpy
h Specific (or molar) enthalpy
h^* Enthalpy of air water vapor mixutre
h_f Specific enthalpy of saturated liquid
h_{fg} Latent heat of vaporization
h_g Specific enthalpy of saturated vapor
J Joule
K Kelvin
K Spring constant
KE Kinetic energy
l Length
M Molar mass
m Mass
m Metre
m_f Mass of liquid
m_g Mass of vapor
N Newton
n Number of moles
P Pressure
PE Potential energy
Pa Pascal
P^s Saturation pressure
P_c Critical pressure
P_m Mean effective pressure
P_R Reduced pressure
P_{tp} Pressure at triple point of water
p_a Partial pressure of air
p_w Partial pressure of water vapor
Q Energy transferred as heat
Q Rate of heat flow
R Specific gas constant
R_u Universal gas constant
r_c Cutoff ratio
r_0 Compression ratio
r_p Pressure ratio of Brayton cycle
S Entropy
s Specific (or molar) entropy
s Second
s_f Specific entropy of saturated liquid
s_g Specific entropy of saturated vapor
T Temperature
T_c Critical temperature
T_R Reduced temperature
t Time
U Internal energy
u Specific (or molar) internal energy
V Volume
v Specific (or molar) volume
v_c Critical volume
v_f Specific volume of saturated liquid
v_g Specific volume of saturated vapor
v_R Reduced volume
v'_R Pseudo-reduced volume
$\mathbf{V}$ Velocity
W Work done
W Watt
W_s Shaft work done
X Quality
Z Compressibility factor
z Elevation

Greek Letters

γ Ratio of specific heats
η Thermal efficiency
η_P Isentropic efficiency of a pump
η_T Isentropic efficiency of a turbine
κ Isothermal compressibility
μ_{JT} Joule–Thomson coefficient
ρ Density
ϕ Relative humidity
ω Specific humidity

1 Fundamental Concepts and Definitions

1.1 What is Thermodynamics?

Thermodynamics is the science dealing with energy and its transformation. It deals with the relationships between heat and work, and the properties of a system in equilibrium. Historically, thermodynamics originated as a result of man's endeavour to convert heat into work. Thermodynamics was formalized in the nineteenth century by Joule (1818–1889), Kelvin (1824–1907), Clausius (1822–1888) and Carnot (1792–1932). Later Gibbs (1839–1903) developed and widened its scope in such a way that nowadays thermodynamics is extensively used in all branches of engineering and physical sciences.

The principles of thermodynamics are summarized in the form of four laws known as the zeroth, first, second and third laws of thermodynamics. The Zeroth Law of Thermodynamics deals with thermal equilibrium and provides a means for measuring temperature. The First Law of Thermodynamics deals with the conservation of energy and introduces the concept of internal energy. The Second Law of Thermodynamics dictates the limits on the conversion of heat into work and provides a yardstick to measure the performance of various processes. It also tells whether a particular change or process is feasible or not, specifies the direction in which the processes will proceed and introduces the concept of entropy. Finally, the Third Law of Thermodynamics defines the absolute zero of entropy. The laws of thermodynamics cannot be proved directly as we prove theorems in mathematics but their validity can be judged by verifying the consequences predicted by the application of these laws to physical and chemical processes. These laws of thermodynamics are based on logical reasoning and were deduced from experimental observations. So far no experimental evidence is available to doubt their validity.

1.2 Macroscopic and Microscopic Approaches

In the study of thermodynamics one can adopt two different points of view, namely macroscopic and microscopic. Suppose a certain quantity of a gas is contained in a closed vessel. One can measure the volume occupied by the gas by measuring the dimensions of the vessel. One can also measure the pressure exerted by the gas with the help of a pressure gauge and measure the temperature of the gas with the help of a thermometer or a thermocouple. The state or condition of the gas can be described by specifying the pressure (P), temperature (T) and volume (V) of the gas.

In the macroscopic approach, the state of the matter under consideration is specified by variables like P, T and V. The variables which are used to describe the state of the matter, can be easily measured or estimated. In this approach, the structure of the matter is not considered, i.e., one need not specify whether the crystalline structure is a body-centred cubic or a face-centred cubic, etc. Classical thermodynamics adopts the macroscopic approach.

One can also view the gas contained in the vessel as a collection of a large number of molecules, each of which is moving at random with independent velocity. The gas contained in the vessel can be described by specifying the position coordinates (x, y, z) and velocity coordinates $(\mathrm{V}_x, \mathrm{V}_y, \mathrm{V}_z)$ of each molecule at a given instant of time. The position and velocity coordinates of the molecules keep changing with time due to collisions between the molecules. Such a description of the state of matter which constitutes the essence of the microscopic approach requires a large number of variables and none of these variables can be measured. Since the properties of the matter depend on these variables ($x, y, z, \mathbf{V}_x, \mathbf{V}_y, \mathbf{V}_z$) which also change with time, one has to obtain the time-average values (i.e., the average value over a long period of time) of these properties. To deduce the time-average values of the properties it is essential to use some kind of statistical averaging. The type of statistics to be employed depends on the nature of the matter under consideration. The microscopic approach is adopted in statistical thermodynamics. In this approach a knowledge of the structure of the matter under consideration is essential and a large number of variables, which cannot be directly measured, are needed to describe the state of matter.

The purpose of this book is to serve as an introduction to the study of thermodynamics and hence we adopt the macroscopic approach only. For better understanding of thermodynamics, it is essential to have a sound knowledge of the units of measurement. Although the system of units employed in thermodynamics is identical with that employed in the physical sciences, a brief review of the units is presented below.

1.3 SI Units

All physical quantities are divided into two groups—primary quantities and secondary quantities. The primary quantities are expressed in terms of base units and the secondary quantities are expressed in terms of derived units. The International System of units or SI units will be followed throughout this book. The basic units of mass, length, time and temperature in the SI units are given below.

1.3.1 Kilogram

The basic SI unit of mass is the kilogram (abbreviated as kg). A standard block of platinum-iridium alloy maintained at the International Bureau of Weights and Measures at Sevres, Paris is taken as the basic unit of mass.

1.3.2 Mole

We know that all materials consist of atoms, molecules, etc. Therefore, the quantity of a substance can also be expressed in terms of the number of atoms (or molecules) present in the matter. Quite often the quantity of matter is expressed in terms of the unit mole (abbreviated as mol). The mole is the amount of substance which contains as many atoms (or molecules) as there are atoms in 0.012 kg of carbon-12.

1.3.3 Metre

The basic SI unit of length is the metre (abbreviated as m). The metre is defined as the length equal to 1,650,763.73 wavelengths in vacuum of the radiation corresponding to the transition between the levels $2p_{10}$ and $5d_5$ of the krypton-86 atom.

1.3.4 Second

The basic SI unit of time is the second (abbreviated as s). The second is defined as the duration of 9,192,631,770 periods of the radiation corresponding to the transition between the two hyperfine levels of the ground state of the cesium-133 atom.

1.3.5 Kelvin

The basic SI unit of temperature is the kelvin (abbreviated as K). The kelvin is defined as the fraction 1/273.16 of the thermodynamic temperature of the triple point of water.

Quite often, the Celsius temperature (symbol °C) is used to express the temperature of a substance. The relation between Kelvin and Celsius temperature is given by K = °C + 273.15.

1.3.6 Derived Units

The secondary quantities are expressed in terms of derived units. The relation between derived units and base units depends on a law or definition. For example, the velocity V is defined as $V = dl/dt$, where l is the length and t is the time. Similarly, the acceleration a is defined as $a = dV/dt$. Therefore, the units of velocity and acceleration are m/s (or m s^{-1}) and m/s^2 (or m s^{-2}), respectively. The units of force are derived from Newton's second law of motion given by $F = ma$, where F is the force. The units of force are kg m/s^2. In honor of Newton, the SI unit of force is named as newton (abbreviated as N). That is, one unit of newton is the force required to produce an acceleration of 1 m/s^2 in a body of mass 1 kg. The SI units of pressure, energy and power are as given below.

$$\text{Pressure} = \frac{\text{Force}}{\text{Area}} = \frac{\text{N}}{\text{m}^2} \text{ or } (\text{N m}^{-2}) = \text{Pa} = \text{pascal}$$

$$\text{Energy} = \text{Force} \times \text{Distance} = \text{N m} = \text{J} = \text{joule}$$

$$\text{Power} = \frac{\text{Energy}}{\text{Time}} = \frac{\text{J}}{\text{s}} \text{ or } (\text{J s}^{-1}) = \text{W} = \text{watt}$$

The SI units of pressure, energy and power are named in honor of the scientists Pascal, Joule and Watt, respectively. It was internationally agreed to abbreviate the units named after the scientists by capital letters, and all the other units are abbreviated using lower case letters. The dimensions and units which are in wide use in thermodynamics are presented in Table 1.1.

Table 1.1 *Dimensions and units of a few physical quantities*

Quantity	Dimensions	Units	Abbreviation	Name
Mass	M	kilogram	kg	
		mole	mol	
Length	L	metre	m	
Time	t	second	s	
Temperature	T	kelvin	K	kelvin
Velocity	Lt^{-1}	m/s		
Acceleration	Lt^{-2}	m/s^2		
Force	MLt^{-2}	kg m/s^2	N	newton
Pressure	$ML^{-1}t^{-2}$	kg/m s^2 or (N/m^2)	Pa	pascal
Energy	ML^2t^{-2}	kg m^2/s^2 (or N m)	J	joule
Power	ML^2t^{-3}	kg m^2/s^3 (or J/s)	W	watt

In order to express very small or very large quantities, the SI units can be expressed in terms of appropriate multipliers. The commonly used prefixes and their symbols which can be combined with the SI units are given in Table 1.2.

Table 1.2 SI unit prefixes

Factor	Prefix	Symbol	Factor	Prefix	Symbol
10^3	kilo	k	10^{-3}	milli	m
10^6	mega	M	10^{-6}	micro	μ
10^9	giga	G	10^{-9}	nano	n
10^{12}	tera	T	10^{-12}	pico	p
10^{15}	peta	P	10^{-15}	femto	f
10^{18}	exa	E	10^{-18}	atto	a

The SI unit of pressure (pascal) is very small in magnitude. For convenience the pressure is usually expressed in kilopascals (abbreviated as kPa) or megapascals (abbreviated as MPa). The standard atmospheric pressure (1 atm) is equal to 101.325 kPa. The units of pressure which are in common use are

$$1 \text{ bar} = 10^5 \text{ Pa} = 100 \text{ kPa} = 0.1 \text{ MPa}$$

$$1 \text{ atm} = 101.325 \text{ kPa} = 0.981 \text{ bar} = 0.981 \times 10^5 \text{ Pa}$$

The pressure exerted on a boundary wall is called the *absolute pressure*. The pressure exerted by a gas in a tank is usually measured with the help of a pressure gauge for which the reference is atmospheric pressure. The pressure recorded by a pressure gauge is called *gauge pressure* which is equal to the difference between the absolute pressure and atmospheric pressure. Vacuum represents the negative gauge pressure or the difference between the atmospheric pressure and the absolute pressure. That is, if a pressure gauge reads the pressure as 150 kPa, then the absolute pressure is equal to 150 + 101.325 = 251.325 kPa. Similarly, if a vacuum gauge reads 50 kPa, then the absolute pressure is given by

$$50 \text{ kPa} = 101.325 \text{ kPa} - P \quad \text{or} \quad P = 101.325 - 50 = 51.325 \text{ kPa}$$

1.4 Conventions used in SI Units

The following conventions are recommended by the International Committee of Weights and Measures to express the SI units and quantities.

- The symbol of a prefix should be combined with a single unit symbol to form a new symbol. For example, 1000 × Pa = kPa.
- Compound prefixes should not be used to express a quantity. For example, 10^{-9} m should be written as nm and not as mμm.
- The SI base unit of mass is kg which already contains the prefix k. Therefore, 10^{-6} kg should be written as 1 mg and not as 1 μkg.
- While writing SI units which are named after scientists in full, only lower case letters

should be used. For example, the unit of pressure should be written as pascal and not as Pascal.

- The unit symbols should not be used in plural form. For example, 100 joules should be written as 100 wJ and not as 100 Js.
- A small gap should be left in between the numerical value and the unit symbol. For example, 100 joules should be written as 100 J.
- When a unit is obtained by the multiplication of two or more units they should be separated by a small gap and should not indicate the multiplier sign. For example, newton × second should be written as N s but not as N × s or N·s.
- The SI prefixes should be selected such that the numerical value lies between 0.1 and 1000. For example, 100,000 Pa can be expressed as 100 kPa or 0.1 MPa. However, when the values are presented in a tabular form, it is advisable to use the same prefix throughout even if some of the values lie outside the range of 0.1 to 10000.

EXAMPLE 1.1 Express the following quantities in proper SI units. (a) 150,000,000 watts (b) 5 angstroms (c) 1 gram/cubic centimetre (d) 0.0025 joules (e) 12 grams per minute.

SOLUTION (a) 150,000,000 watts $= 1.5 \times 10^8$ watts. It can be expressed as 150 MW or 0.15 GW

(b) 5 angstroms $= 5 \times 10^{-8}$ cm $= 5 \times 10^{-10}$ metres $= 0.5 \times 10^{-9}$ metres. It can be correctly expressed as 0.5 nm

(c) $$\frac{1 \text{ gram}}{\text{cubic centimetre}} = \frac{1 \times 10^{-3} \text{ kg}}{(1 \times 10^{-2} \text{ metre})^3} = \frac{1 \times 10^{-3}}{10^{-6}} = 1000 \text{ kg/m}^3$$

(d) 0.0025 joules $= 2.5 \times 10^{-3}$ J $= 2.5$ mJ

(e) $$\frac{12 \text{ grams}}{\text{minute}} = \frac{12 \times 10^{-3} \text{ kg}}{60 \text{ seconds}} = 0.2 \times 10^{-3} \text{ kg/s} = 0.2 \text{ g/s}$$

EXAMPLE 1.2 According to Newton's universal law of gravitation, the force (F) between two bodies of masses m_1 and m_2 which are separated by a distance r is given by

$$F = G\frac{m_1 m_2}{r^2}$$

where G is the universal gravitational constant. Determine the SI units of G.

SOLUTION $F = G\dfrac{m_1 m_2}{r^2}$ or $G = \dfrac{Fr^2}{m_1 m_2}$

$$\text{Units of } G = \frac{\text{N} \times \text{m}^2}{(\text{kg})^2} = \text{N m}^2/\text{kg}^2$$

1.5 System

In any scientific or engineering analysis it is essential to clearly identify the material on which we focus our attention. In mechanics, when we want to analyze the motion of a body with the help of Newton's laws of motion, we draw a free body diagram by isolating the body under consideration from its surroundings. For this purpose, we draw an imaginary boundary enclosing the body or a part of it and represent the various forces acting on it. Similarly in thermodynamic analysis we enclose the body under consideration by an imaginary boundary and represent the various energy flows across the boundary. For this purpose we use the term *thermodynamic system* or simply *system* to refer to a definite quantity of matter on which we focus our attention and which is enclosed by a boundary. The boundary may be imaginary or real. Suppose we want to analyze the energy changes of the air in a room, we may focus our attention on 1 m^3 of air in the room and enclose it in an imaginary boundary. On the other hand, if we are interested in analyzing the energy changes of a gas enclosed in a piston-cylinder assembly while it is compressed or expanded, we can draw a boundary enclosing the gas and the gas in the enclosure is our system. In this case, the boundary coincides with the physical boundary of the cylinder and the volume of the system can change. Energy in the form of heat and work can cross the boundary of the system. A few systems are shown in Fig. 1.1.

A system refers to a definite quantity of matter. That is, the mass of the system remains unchanged as it undergoes certain processes. The system may exchange heat and work with other systems or bodies. Sometimes, a system is also called a control mass or closed system. A system which is enclosed by an adiabatic wall, i.e., the boundary which does not allow energy exchange in the form of heat is called an *adiabatic system*. An adiabatic system can exchange energy in the form of work with the surroundings or with other bodies. A system which does not exchange energy in the form of heat as well as work with the surroundings or with other bodies is called an *isolated system*. That is, an isolated system is enclosed by a rigid and insulated boundary. Figure 1.2 shows an adiabatic system and an isolated system.

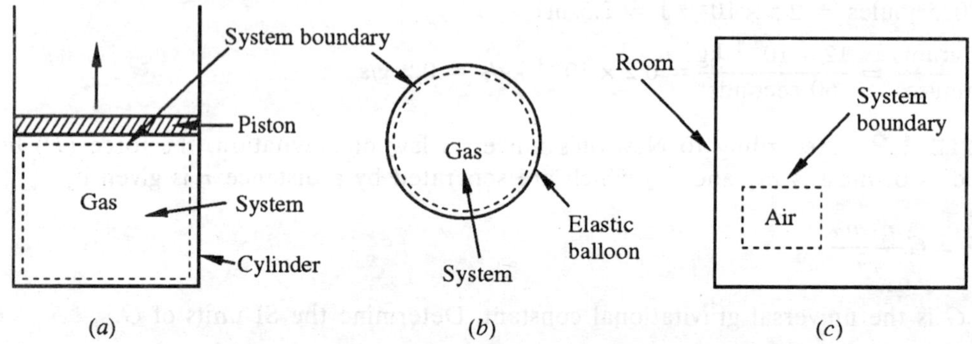

Fig. 1.1 *(a) Gas enclosed in a piston-cylinder assembly. The volume of the system may change though the shape may remain the same. The system boundary coincides with the physical boundaries of the cylinder and it is real. (b) Gas enclosed in an elastic balloon. The volume as well as the shape of the system may change if the balloon is squeezed. The system boundary coincides with the walls of the balloon and the boundary is real. (c) Air in a room. The system boundary is imaginary.*

Quite often an engineer has to deal with equipment like turbines, compressors, pumps, etc., through which material continuously flows in and out. In analyzing the performance of such equipment we focus our attention on a specified region of space (say the equipment itself). The chosen region of space is called a *control volume* or sometimes as an *open system*. The adjectives 'closed' or 'open' used with a system are with respect to the mass flow. In an open system, or better called as control volume, mass may enter or leave the system. The boundary enclosing the control volume is called *control surface*. Energy in the form of work and heat can cross the control surface. A few examples of open systems are shown in Fig. 1.3.

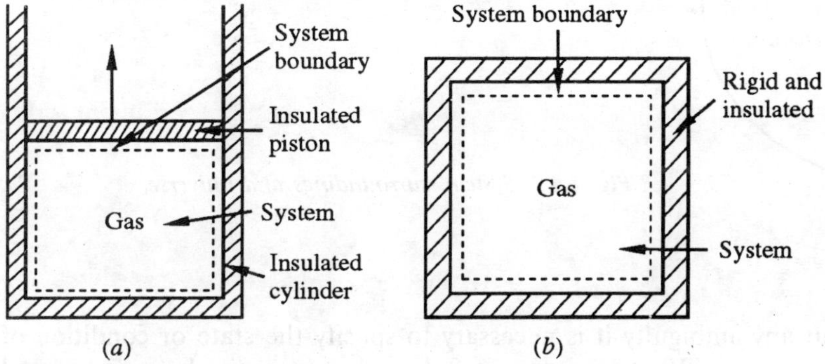

Fig. 1.2 (a) Adiabatic system. (b) Isolated system.

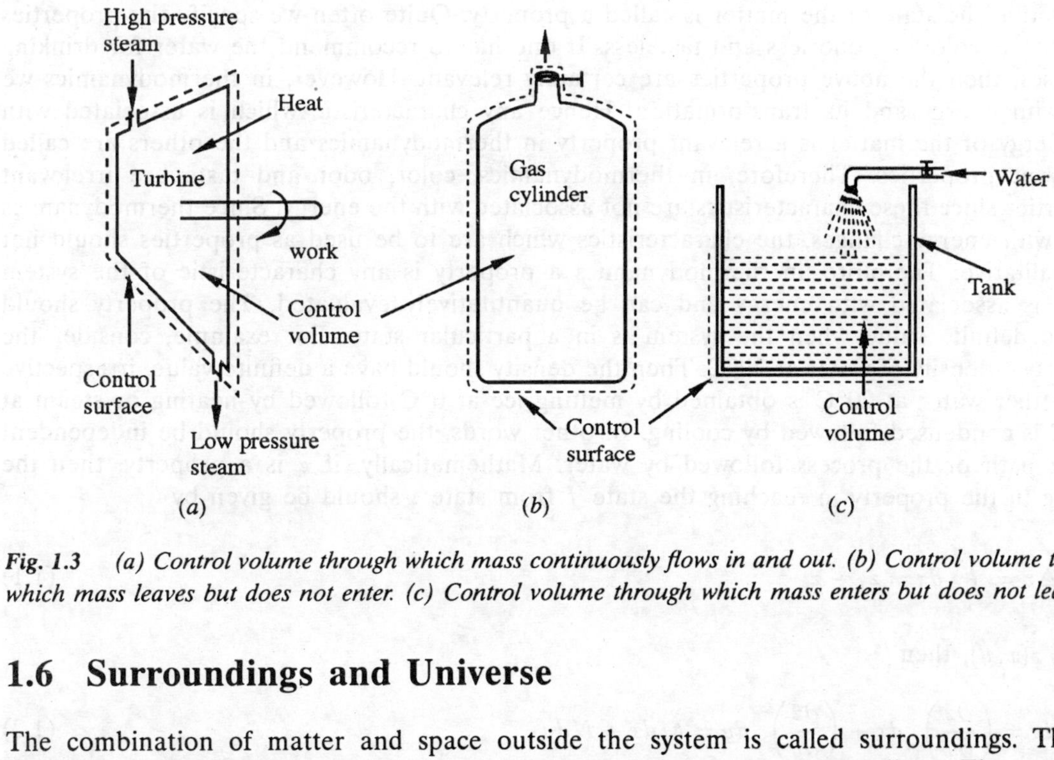

Fig. 1.3 (a) Control volume through which mass continuously flows in and out. (b) Control volume through which mass leaves but does not enter. (c) Control volume through which mass enters but does not leave.

1.6 Surroundings and Universe

The combination of matter and space outside the system is called surroundings. That is, all matter which is not included in the system constitutes the surroundings. The system is separated from the surroundings by the system boundary. The combination of a system and its surroundings is called universe. Figure 1.4 shows a system, surroundings and universe.

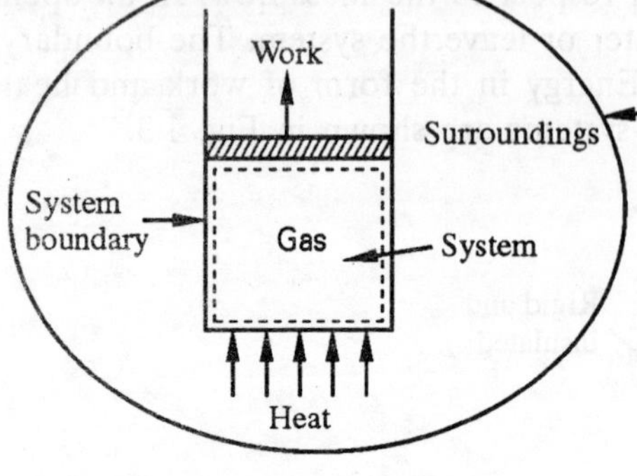

Fig. 1.4 System, surroundings and universe.

1.7 Property

To describe a system without any ambiguity it is necessary to specify the state or condition of the system in terms of its properties. What is a property? Any characteristic that can be used to describe the state of the matter is called a property. Quite often we specify the properties of water as colorless, odorless and tasteless. If one has to recommend the water for drinking purposes, then the above properties are certainly relevant. However, in thermodynamics we deal with energy and its transformation. Hence any characteristic which is associated with the energy of the matter is a relevant property in thermodynamics and the others are called irrelevant properties. Therefore, in thermodynamics, color, odor and taste are irrelevant properties since these characteristics are not associated with the energy. Since thermodynamics deals with energy changes, the characteristics which are to be used as properties should not be qualitative. Therefore, in thermodynamics a property is any characteristic of the system which is associated with energy and can be quantitatively evaluated. The property should have a definite value when the system is in a particular state. For example, consider the property—density of water at 20°C. Then the density should have a definite value, irrespective of whether water at 20°C is obtained by melting ice at 0°C followed by heating or steam at 100°C is condensed followed by cooling. In other words, the property should be independent of the path or the process followed by water. Mathematically, if z is a property, then the change in the property in reaching the state f from state i should be given by

$$\Delta z = \int_i^f dz = z_f - z_i \tag{1.1}$$

If $z = z(x, y)$, then

$$dz = \left(\frac{\partial z}{\partial x}\right)_y dx + \left(\frac{\partial z}{\partial y}\right)_x dy = M\,dx + N\,dy \tag{1.2}$$

where $M = \left(\frac{\partial z}{\partial x}\right)_y$ and $N = \left(\frac{\partial z}{\partial y}\right)_x$.

If $(\partial M/\partial y)_x = (\partial N/\partial x)_y$, then dz is said to be an exact differential and z is a point function.

Therefore, a thermodynamic property is a point function and not a path function. The characteristics which satisfy the above criterion are pressure (P), temperature (T), volume (V), density (ρ), etc., and hence they are thermodynamic properties of a system. The density (ρ) is mass per unit volume or the density is defined as

$$\rho = \lim_{dV \to 0} \left(\frac{dm}{dV} \right) = \text{kg/m}^3 \tag{1.3}$$

where dm is the mass of a system in a differential volume dV.

If the number of molecules in a unit volume do not change from point to point in a system, then every unit volume will contain the same amount of mass. In such cases, we can write $\rho = m/V$. If the above condition is satisfied, then the system is considered to be a continuum.

EXAMPLE 1.3 For a particular body, the differential of the volume V is given by $dV = 2\pi rhdr + \pi r^2 dh$. Test whether the quantity V can be used as a thermodynamic property or not. If it can be used, determine the relation $V = V(r, h)$.

SOLUTION The given differential is $dV = 2\pi rhdr + \pi r^2 dh = Mdr + Ndh$.

If the given differential is exact, then it can be used as a property. Therefore, let us check whether dV is an exact differential or not. If it is an exact differential, it should satisfy the relation

$$\left(\frac{\partial M}{\partial h} \right)_r = \left(\frac{\partial N}{\partial r} \right)_h \tag{A}$$

$$\left(\frac{\partial M}{\partial h} \right)_r = \left[\frac{\partial}{\partial h}(2\pi rh) \right]_r = 2\pi r$$

$$\left(\frac{\partial N}{\partial r} \right)_h = \left[\frac{\partial}{\partial r}(\pi r^2) \right]_h = 2\pi r$$

Since Eq. (A) is satisfied, V can be used as a thermodynamic property. To obtain the relation $V = V(r, h)$, we can partially integrate M with respect to r or N with respect to h.

$$V = \left[\int M dr \right]_h = \left[\int 2\pi rhdr \right]_h = \pi r^2 h + f(h) \tag{B}$$

To determine the constant $f(h)$, let us evaluate $(\partial V/\partial h)_r$ and compare it with N.

$$\left(\frac{\partial V}{\partial h} \right)_r = \left[\frac{\partial}{\partial h}[\pi r^2 h + f(h)] \right]_r = \pi r^2 + \frac{df(h)}{dh} = N = \pi r^2$$

or $\dfrac{df(h)}{dh} = 0$ or $f(h) = \text{constant}$

Therefore $V = \pi r^2 h +$ constant.

1.8 Intensive and Extensive Properties

Consider the systems A and B shown in Fig. 1.5. The volume of system A is 1 m^3 while the volume of system B is 2 m^3. If we consider the combination of systems A and B as a system C, then we find the volume of system C is 3 m^3. That is, the property volume depends on the size or mass of the system under consideration. On the other hand, the properties temperature and pressure are the same for all the three systems A, B and C and they do not depend on the mass of the system under consideration. Properties like volume which depend on the mass of a system are called *extensive properties* and properties like pressure and temperature which do not depend on the mass of the system are called *intensive properties.*

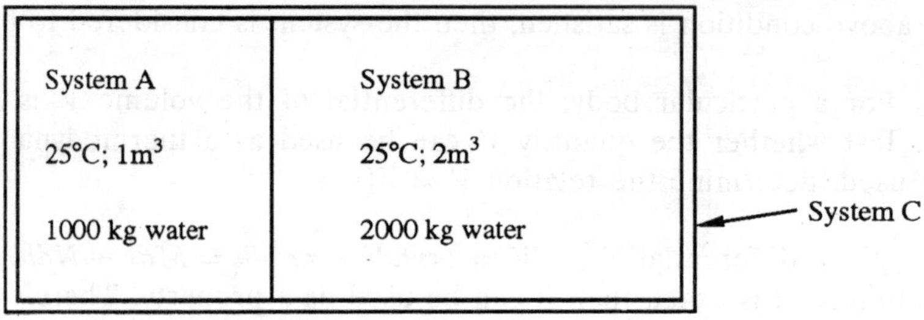

Fig. 1.5 Extensive and intensive properties of a system.

The ratio of the extensive property to the mass is called the *specific property.* For example, the specific volume v is given by

$$v = V/m = \text{m}^3/\text{kg} \tag{1.4}$$

It can be observed from Eqs. (1.3) and (1.4) that the density is equal to the inverse of specific volume. That is,

$$\rho = 1/v \tag{1.5}$$

The ratio of the extensive property to the moles of the system is called molar property. For example, the molar volume, v is given by

$$v = V/n = \text{m}^3/\text{mol} \tag{1.6}$$

1.9 Phase

Substances can be found in three different states of physical aggregation. For example, ice, water and steam are the three different states of the same chemical species H_2O. The different states in which a pure substance can exist are called its phases. If a system consists of a mixture of phases (for example, a mixture of ice and water), the phases are separated from each other by a phase boundary. The thermodynamic properties of the substance change abruptly at the phase boundary, even though the intensive properties like pressure and temperature are identical as shown in Fig. 1.6.

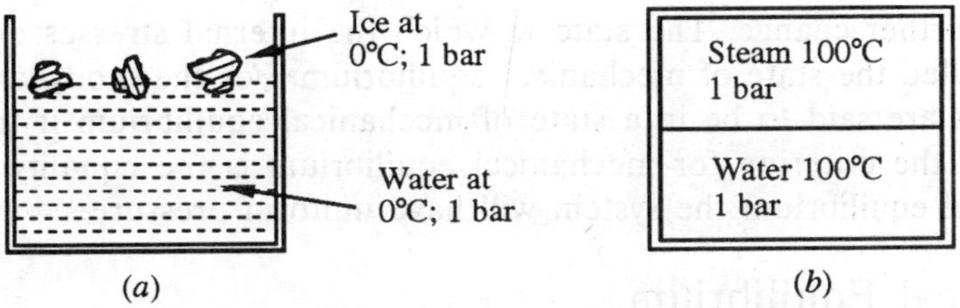

Fig. 1.6 *(a) Mixture of ice and water at* 0°C *and* 100 kPa. *The density of ice is different from that of liquid water. (b) Steam and liquid water at* 100°C *and* 100 kPa. *The specific volume of steam is different from that of liquid water.*

Mixtures also exist in different phases. For example, a mixture of alcohol and water can be present in liquid as well as vapor phases. The chemical composition of the vapor phase will be usually different from the chemical composition of the liquid phase. Solids can exist in different allotropic forms. For example, sulfur can be found in rhombic form (lemon yellow crystals) as well as in monoclinic form (orange crystals). The properties of these allotropes differ from each other. Similarly carbon can be found in diamond form as well as in graphite form. Now we can generalize the above observations and define a phase as a system which is uniform throughout, both in the chemical composition and the physical state.

1.10 Equilibrium

Equilibrium is a concept associated with the absence of any tendency for spontaneous changes when the system is isolated. Once the system reaches a state of equilibrium all macroscopically measurable properties remain uniform and constant, even though there will be changes at the microscopic level. There are different types of equilibrium, namely thermal, mechanical and chemical equilibrium.

1.10.1 Thermal Equilibrium

Consider two blocks of metal, one hot and the other cold. If the two blocks are brought into physical contact with each other, we observe that the temperature of the cold block keeps increasing while the temperature of the hot block keeps decreasing. The temperatures of both the blocks keep changing with time. If we continuously monitor the temperatures of both the blocks, after sometime we find that the temperatures of both the blocks will be identical and no further change occurs. At this state, the two blocks are said to be in a state of thermal equilibrium. Therefore, we can say that two given systems are in a state of thermal equilibrium if they have the same temperature. That is, the criterion for thermal equilibrium is the equality of temperature. In a state of thermal equilibrium, the system will have a uniform temperature.

1.10.2 Mechanical Equilibrium

If a force is applied to a system, internal stresses will develop within the system to balance the applied force and after sometime, the stresses will exactly balance the applied force and there will be no further change. The state in which the internal stresses exactly balance the applied force is called the state of mechanical equilibrium. For thermodynamic purposes, two interacting systems are said to be in a state of mechanical equilibrium if their pressures are the same. That is, the criterion for mechanical equilibrium is the equality of pressure. In a state of mechanical equilibrium the system will have uniform pressure.

1.10.3 Chemical Equilibrium

Suppose a small crystal of potassium permanganate is dropped in a beaker containing water. Immediately after dropping the crystal, the color of water in the immediate vicinity of the crystal will be dark purple and at other places the water will be colorless. As time progresses we find that the color of water in the top layer slowly changes from colorless to light purple color. If we wait for a sufficient period of time all the water in the beaker will have a uniform color. The intensity of color is related to the concentration of potassium permanganate in water. The potassium permanganate diffuses from the region of high concentration to the region of low concentration. When the concentration is uniform, diffusion or mass transfer ceases and in that state, the system is said to be in a state of chemical equilibrium. In a homogeneous or single phase system, if the concentration is uniform and there is no tendency for mass transfer or diffusion, the system is said to be in a state of chemical equilibrium. The criterion for chemical equilibrium is the equality of chemical potential. The chemical potential is not discussed here as it is beyond the scope of this text. A system which is simultaneously in a state of thermal, mechanical and chemical equilibrium is said to be in a state of *thermodynamic equilibrium*.

1.11 Process

In thermodynamics, we are concerned with systems in equilibrium. A process is the path followed by a system in reaching a given final equilibrium state starting from a specified initial state. Suppose a certain quantity of ideal gas is held in a piston-cylinder assembly at pressure P_1 and temperature T as shown in Fig. 1.7. If the energy is transferred as heat to the gas while its temperature is held constant, the gas expands till its pressure reduces to P_2. Then we say that the system (gas) has undergone a process at constant temperature.

Quite often we shall be dealing with the following processes:

- Constant temperature (or isothermal) process.
- Constant pressure (or isobaric) process.
- Constant volume (or isochoric, or isometric) process.
- Adiabatic process.

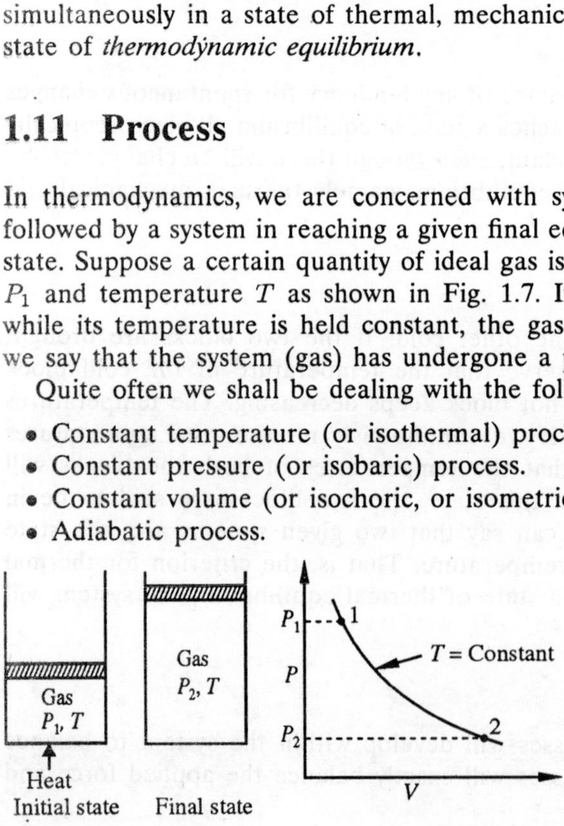

Fig. 1.7 Constant temperature process.

Suppose a system undergoes a series of processes in such a way that the system is finally restored to its initial state, then the system is said to have undergone a cyclic process. A few examples of cyclic processes are shown in Fig. 1.8.

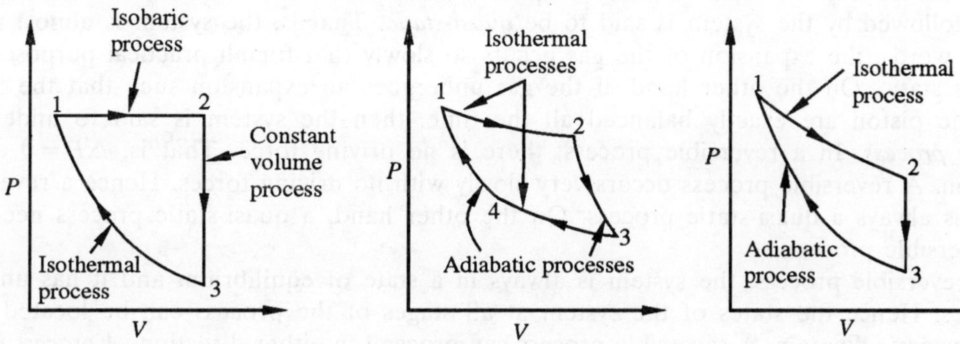

Fig. 1.8 Examples of cyclic processes.

1.12 Reversible Process

Suppose a certain quantity of gas is enclosed in a piston-cylinder assembly and the piston is loaded with a heap of fine sand as shown in Fig. 1.9. Initially, the pressure of gas in the

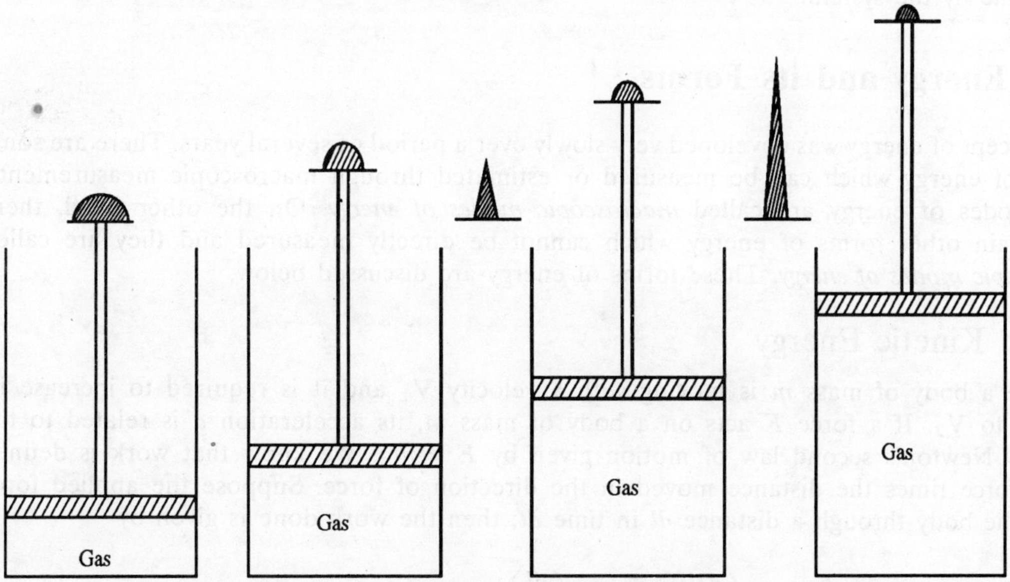

Fig. 1.9 Example of a quasi-static process.

cylinder is exactly balanced by the pressure exerted by the sand. If a particle of sand (of negligible mass) is removed from the piston top, the downward pressure exerted by the sand is infinitesimally smaller than the gas pressure and the gas expands. By continuing the process of removing the sand particles one after another, the gas can be made to expand in such a way that at any given instant of time, the gas pressure is almost exactly balanced by the pressure exerted by the sand. In fact the pressure difference ΔP is infinitesimal. Then the process followed by the system is said to be *quasi-static*. That is, the system is almost static. In other words, the expansion of the gas occurs so slowly that for all practical purposes, the system is static. On the other hand, if the gas undergoes an expansion such that the forces across the piston are exactly balanced all the time, then the system is said to undergo a *reversible process*. In a reversible process, there is no driving force. That is, $\Delta P = 0$ across the piston. A reversible process occurs very slowly with no driving forces. Hence a reversible process is always a quasi-static process. On the other hand, a quasi-static process need not be a reversible process.

In a reversible process, the system is always in a state of equilibrium and it has uniform properties. Hence the states of the system at all stages of the process can be located on a thermodynamic diagram. A reversible process can proceed in either direction. *A process is said to be reversible if both the system and its surroundings can be restored to their respective initial states by reversing the direction of the process.* A process is irreversible if it does not satisfy this criterion. Expansion or compression with no pressure difference across the system boundary are examples of reversible processes.

All actual processes are irreversible and non-quasi-static because they proceed with finite pressure or temperature differences and they occur at a finite rate. For thermodynamic analysis we idealize several processes as reversible because reversible processes give the maximum work done by the system.

1.13 Energy and its Forms

The concept of energy was developed very slowly over a period of several years. There are some modes of energy which can be measured or estimated through macroscopic measurements. Such modes of energy are called *macroscopic modes of energy*. On the other hand, there are certain other forms of energy which cannot be directly measured and they are called *microscopic modes of energy*. These forms of energy are discussed below.

1.13.1 Kinetic Energy

Suppose a body of mass m is moving with a velocity $\mathbf{V}_1$ and it is required to increase its velocity to $\mathbf{V}_2$. If a force F acts on a body of mass m, its acceleration a is related to the force by Newton's second law of motion given by $F = ma$. We know that work is defined as the force times the distance moved in the direction of force. Suppose the applied force moves the body through a distance dl in time dt; then the work done is given by

$$dw = F\,dl = (ma)dl = m\left(\frac{d\mathbf{V}}{dt}\right)dl = m\left(\frac{dl}{dt}\right)d\mathbf{V}$$

or $$\int_1^2 dW = \int_1^2 m\mathbf{V}d\mathbf{V} \quad \text{or} \quad W = m(\mathbf{V}_2^2 - \mathbf{V}_1^2)/2 = \Delta\left(\frac{m\mathbf{V}^2}{2}\right)$$

The quantity ($m\mathbf{V}^2/2$) is called *kinetic energy* (KE). The kinetic energy possessed by a body can be measured by measuring its velocity **V** and it is a macroscopic mode of energy. The kinetic energy is given by

$$\text{KE} = \frac{1}{2}m\mathbf{V}^2 \tag{1.7}$$

EXAMPLE 1.4 A car of mass 700 kg is moving with a speed of 80 km/h. Calculate the kinetic energy of the car.

SOLUTION The velocity of the car is given by

$$\mathbf{V} = \frac{80 \times 10^3}{3600} = 22.22 \text{ m/s}$$

The kinetic energy is given by

$$\text{KE} = \frac{1}{2}m\mathbf{V}^2 = \frac{1}{2} \times 700 \times (22.22)^2 = 17.2805 \text{ kJ}$$

EXAMPLE 1.5 A car of mass 700 kg is moving with a speed of 20 km/h and it is required to accelerate it to a speed of 100 km/h. Calculate the work to be done on the car.

SOLUTION Initial speed $\mathbf{V}_1 = \dfrac{20 \times 10^3}{3600} = 5.56$ m/s

Final speed $\mathbf{V}_2 = \dfrac{100 \times 10^3}{3600} = 27.78$ m/s

Work to be done is given by

$$W = \frac{1}{2}m(\mathbf{V}_2^2 - \mathbf{V}_1^2) = \frac{1}{2} \times 700 \times (27.78^2 - 5.56^2) = 259.285 \text{ kJ}$$

1.13.2 Potential Energy

If a body of mass m is moved from an initial elevation z_1 to a final elevation z_2, the work done on the body is given by

$$\int dW = \int F dz = \int_1^2 (mg)dz = mg(z_2 - z_1) = \Delta(mgz)$$

The quantity (mgz) is called *potential energy*. That is

$$\text{PE} = mgz \tag{1.8}$$

The potential energy possessed by a body can be estimated by measuring its elevation from a datum plane and hence it is another macroscopic mode of energy.

EXAMPLE 1.6 An elevator of mass 1000 kg is taken from the ground floor to the 21st floor which is 63 m above the ground floor. Calculate the work done on the elevator.

SOLUTION Given: $z_1 = 0$; $z_2 = 63$ m

Work done, $W = mg(z_2 - z_1) = 1000 \times 9.81 \times (63 - 0) = 618.03$ kJ

The sum of the kinetic energy and potential energy possessed by a body is called the *mechanical energy*. The law of conservation of energy for mechanical systems states that the mechanical energy of an isolated mechanical system remains constant. That is, for an isolated mechanical system, the sum of KE and PE remains constant and the energy can transform from KE to PE or vice versa.

EXAMPLE 1.7 In a particular fountain, water is ejected from the tip of a nozzle at a velocity of 10 m/s. Estimate the height to which the water can reach.

SOLUTION KE of water $= (1/2) \times 1 \times (10)^2 = 50$ J/kg

All this KE can be converted to PE and at that instant, the velocity of water will be zero. That is, the maximum PE possessed by water is 50 J/kg. Then

$$\text{PE} = 50 = mgz = 1 \times 9.81 \times z \quad \text{or} \quad z = 5.1 \text{ m}$$

Hence, the water can reach a height of 5.1 m.

1.13.3 Internal Energy

We know that all matter consists of a large number of atoms or molecules and these atoms/molecules are in continuous motion. Associated with the motion of the molecules there are several forms of energy. Consider a diatomic molecule like N_2. The molecule as a whole can move in x, y and z directions with velocities $\mathbf{V}_x, \mathbf{V}_y$ and $\mathbf{V}_z$ and hence it possesses kinetic energy. The kinetic energy associated with the translational motion of the molecules is called *translational energy*. The molecule can rotate about its centre of mass and possess kinetic energy which is given by $(1/2)I\omega^2$, where I is the moment of inertia about the centre of mass and ω is the angular velocity. The kinetic energy associated with the rotational motion of the molecules is called *rotational energy*. In addition to these, the bond length also undergoes changes and the energy associated with such motion of the molecule is called *vibrational energy*. The electrons move around the nucleus and they possess a certain energy called *electronic energy*. All these energies cannot be estimated through direct measurement and these forms of energy are called microscopic modes of energy. The various microscopic modes of energy are shown in Fig. 1.10. The microscopic modes of energy are due to the internal structure of

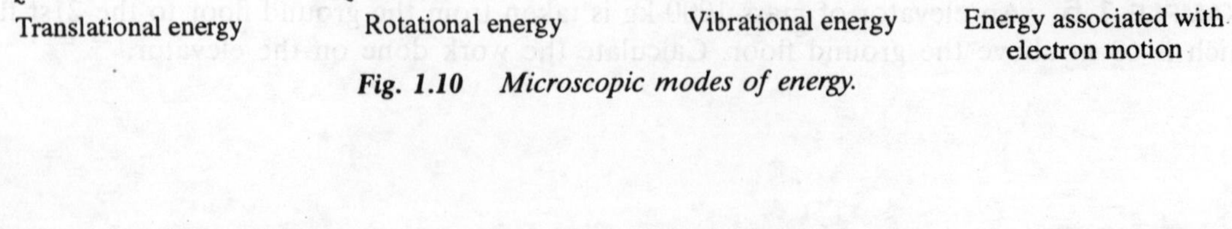

Fig. 1.10 Microscopic modes of energy.

the matter and hence the sum of all the microscopic modes of energy is called the *internal energy* (U). The total energy (E) possessed by a body is given by

$$E = \text{KE} + \text{PE} + U \tag{1.9}$$

1.14 Work

Thermodynamics deals with energy and its transformation. Whenever a system interacts with its surroundings, it can exchange energy with its surroundings in two different ways—Work and Heat. A precise understanding of the terms work and heat is quite essential in thermodynamic analysis of processes.

In mechanics, work is defined as the product of the force and the displacement in the direction of the force. That is, the differential work done (dW) associated with a differential displacement (dl) is given by

$$dW = Fdl \tag{1.10}$$

Let us apply this definition of work to obtain a relation to calculate the work done when a gas undergoes an expansion process. For this, consider the gas enclosed in the piston-cylinder assembly shown in Fig. 1.11. Let A represent the cross-sectional area of the piston and dl the displacement of the piston. Then the differential work done by the gas is given by

$$dW = Fdl = PA(dl) = PdV$$

If the gas is allowed to expand reversibly from the initial pressure P_1 to the final pressure P_2, then the work done is given by

$$W = \int_1^2 PdV$$

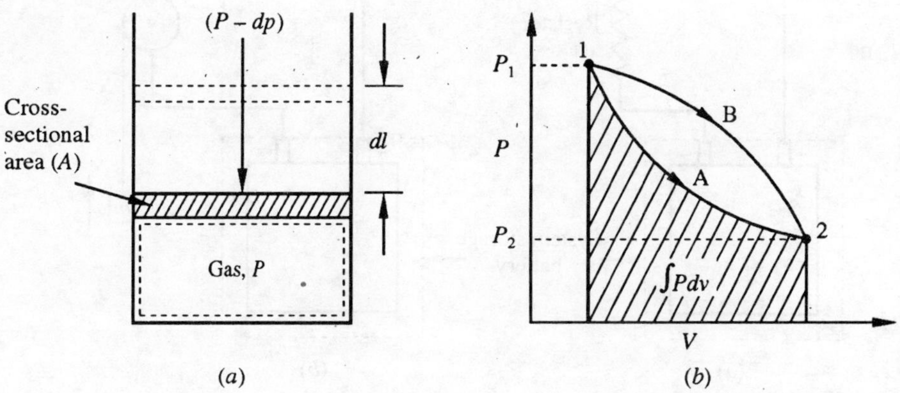

Fig. 1.11 *(a) Work done during an expansion process. (b) $W = \int_1^2 PdV$ and W depends on the path.*

The quantity $\int PdV$ represents the area under the curve on a pressure versus volume diagram as shown in Fig. 1.11(b). If the system follows the path 1-A-2, the work done is given by the shaded area in Fig. 1.11(b). On the other hand, if the system follows the path 1-B-2, the work done is given by the area under the curve 1-B-2, which is different. Therefore, the work depends on the path followed by the system. That is, W is a path function, not a point function. Hence work is not a property of the system.

1.15 Thermodynamic Definition of Work

Consider the storage battery shown in Fig. 1.12 as the system. If the switch is closed, electric current passes through the electrical resistor and the charge of the battery reduces. The system (battery) has interacted with the surroundings. Can we call this interaction as work? In mechanics, this interaction cannot be called work, since no force has moved through a distance. Let us modify the surroundings as shown in Fig. 1.12(b), where the electrical resistor is replaced by a motor which draws the same current as the electrical resistor. One end of a thread is tied to the motor shaft and wound on the shaft, while a mass m is tied to the other end. If the switch is closed, the motor runs and the mass rises through a distance. That is, the battery has done some work in raising a mass. Based on such observation we can provide the thermodynamic definition of work as follows: *Work is said to be done by a system on the surroundings if the sole effect external to the system could be reduced to the raising of a mass through a distance.*

The thermodynamic definition of work specifies that the sole effect external to the system is the raising of a mass. That is, the raising of a mass is the only effect and this effect is external to the system. In other words, one can make any modification in the surroundings and by modifying the surroundings appropriately, if one can visualize the interaction as the raising of a mass and this is the only effect, then the interaction can be called work.

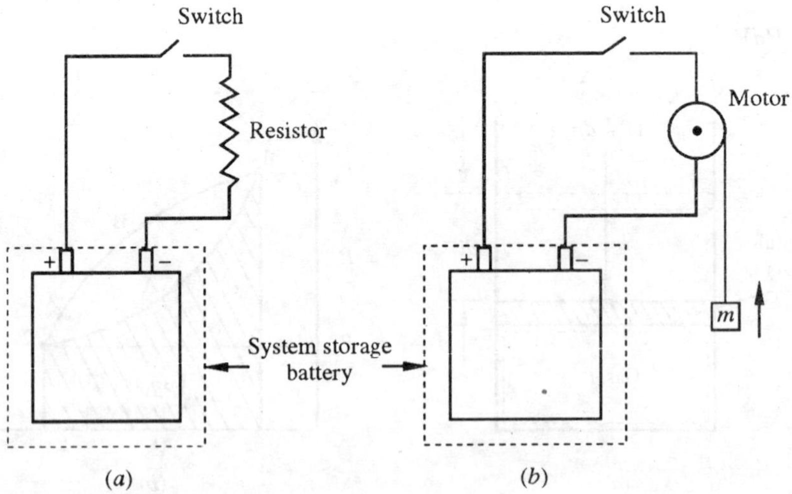

Fig. 1.12 *(a) Interaction of a storage battery with electrical resistor. (b) Visualization of a battery interaction as work.*

Suppose a tank is divided into two compartments by a partition, and one compartment is filled with a gas at pressure P_1 while the second compartment is evacuated. If the partition is punctured, the gas expands and occupies both the compartments. Is there any work done by the gas? Is it equal to $\int PdV$? To determine the area under the curve one can measure the intermediate pressures and volumes by removing the partitions a, b, c, d, e shown in Fig. 1.13 one after another.

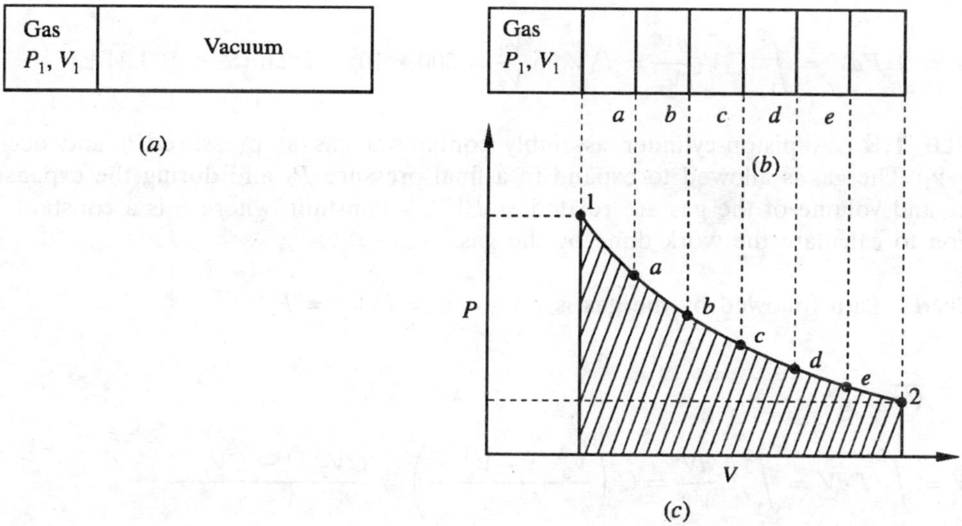

Fig. 1.13 (a) Tank divided into two compartments—one compartment filled with gas at P_1 while the second compartment is evacuated. (b) Several partitions a, b, c, d, e placed in second compartment. (c) P–V diagram showing the intermediate values of P and V as measured after removing the compartments a, b, c, d, e one after another.

By plotting the values of P and V one can obtain the P–V diagram as shown in Fig. 1.13(c). However, the area under the curve is not equal to the work done. In this case we know that the gas has not moved any force through any distance, since the opposing force is equal to zero (because the other side is under vacuum) and hence the work done is equal to zero. The expansion of a gas which is not restrained by an opposing force is called *free expansion*. The work done during free expansion is zero.

$W = \int PdV$ for a reversible process

$W \neq \int PdV$ for an irreversible process

To evaluate $\int PdV$, one should know the relation P and V of the gas for the process under consideration. That is, one should know the path followed by the gas. The calculation of work is illustrated through the following examples.

EXAMPLE 1.8 A piston-cylinder contains a gas at 500 kPa and occupies a volume of 0.2 m^3. Suppose the gas expands to a final pressure of 100 kPa and during expansion the volume of the gas is inversely proportional to pressure. Calculate the work done by the gas.

SOLUTION Given conditions are $P_1 = 500$ kPa; $V_1 = 0.2$ m^3; $P_2 = 100$ kPa

During expansion, $V \propto \dfrac{1}{P}$ or PV = constant = $P_1V_1 = P_2V_2$

or $\dfrac{V_2}{V_1} = \dfrac{P_1}{P_2} = \dfrac{500}{100} = 5$

$$W = \int PdV = \int_1^2 P_1V_1 \frac{dV}{V} = P_1V_1 \ln \frac{V_2}{V_1} = 500 \times 10^3 \times 0.2 \ln(5) = 160.94 \text{ kJ}$$

EXAMPLE 1.9 A piston-cylinder assembly contains a gas at pressure P_1 and occupies a volume V_1. The gas is allowed to expand to a final pressure P_2 and during the expansion the pressure and volume of the gas are related as PV^γ = constant, where γ is a constant. Derive a relation to calculate the work done by the gas.

SOLUTION Path followed by the gas is $PV^\gamma = C = P_1V_1^\gamma = P_2V_2^\gamma$

or $P = \dfrac{C}{V^\gamma}$

$$W = \int_1^2 PdV = \int_1^2 C\frac{dV}{V^\gamma} = C\left(\frac{V_2^{1-\gamma} - V_1^{1-\gamma}}{1-\gamma}\right) = \frac{CV_2^{1-\gamma} - CV_1^{1-\gamma}}{1-\gamma}$$

$$\frac{P_2V_2^\gamma V_2^{1-\gamma} - P_1V_1^\gamma V_1^{1-\gamma}}{1-\gamma} = \frac{P_2V_2 - P_1V_1}{1-\gamma} = \frac{P_1V_1 - P_2V_2}{\gamma - 1}$$

EXAMPLE 1.10 One end of a spring is fixed to a wall and the second end is connected to a block B as shown in Fig. 1.14. Initially the spring is at its natural length. That is, the spring is neither compressed nor elongated. If the block is pulled through a distance l, calculate the work done by the spring. It is given that the force exerted by the spring is proportional to the change in length (that is compression or extension) of the spring.

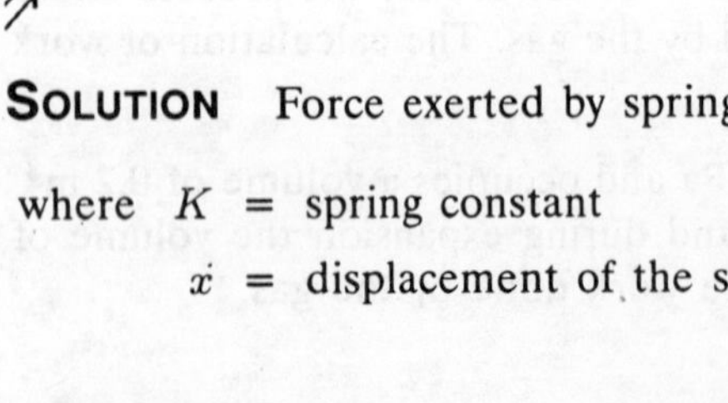

Fig. 1.14 Sketch for Example 1.10.

SOLUTION Force exerted by spring, $F \propto x$ or $F = K \times x$

where K = spring constant

x = displacement of the spring.

Work done, $W = \int_1^2 F dx = \int_1^2 Kx dx = \frac{K}{2}(x_2^2 - x_1^2)$

EXAMPLE 1.11 A particular gas obeys the relation $(P + a/v^2)(v - b) = RT$, where a and b are constants. If such a gas is isothermally expanded from the initial state (P_1, v_1) to the final state (P_2, v_2), determine the work done by the gas.

SOLUTION $(P + a/v^2)(v - b) = RT$ or $P = \frac{RT}{v - b} - \frac{a}{v^2}$

$$W = \int_1^2 P dv = \int_1^2 \left[\frac{RT}{v - b} - \frac{a}{v^2}\right] dv = RT \ln\left(\frac{v_2 - b}{v_1 - b}\right) - a\left(\frac{1}{v_1} - \frac{1}{v_2}\right)$$

EXAMPLE 1.12 A piston-cylinder assembly shown in Fig. 1.15 initially contains a gas at 200 kPa and occupies a volume of 0.1 m^3. The piston is attached to a spring and the force exerted by the spring is proportional to the displacement from its equilibrium position. The atmospheric pressure is 100 kPa. Now, the gas is heated until the volume is doubled and at this stage, the pressure of the gas is 600 kPa. Calculate the work done by the gas.

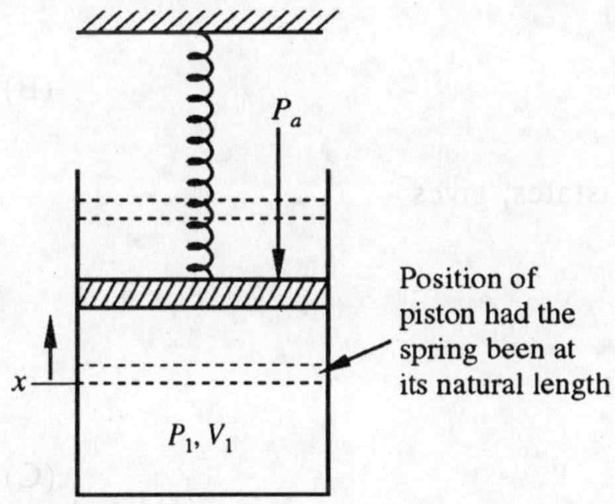

Fig. 1.15 Sketch for Example 1.12.

SOLUTION The given data is $P_1 = 200$ kPa; $V_1 = 0.1$ m^3; $P_2 = 600$ kPa; $V_2 = 0.2$ m^3

Force exerted by spring, $F = Kx$

Let A = cross-sectional area of piston

V_0 = volume of gas if the spring is in its natural length

P_a = atmospheric pressure.

Force balance on the piston at any instant gives

$$PA = P_a A + Kx = P_a A + K\left(\frac{V - V_0}{A}\right)$$

$$P = P_a + \frac{K(V - V_0)}{A^2}$$

Work done, $W = \int_1^2 PdV = \int_1^2 \left[P_a + \frac{K(V - V_0)}{A^2} \right] dV$

or

$$W = P_a(V_2 - V_1) + \frac{K}{2A^2} [(V_2 - V_0)^2 - (V_1 - V_0)^2] \quad \text{(A)}$$

The work done can be calculated from Eq. (A) if we know the values of K, A and V_0. However, this information is not given to us. To evaluate W, Eq. (A) can be rewritten as

$$W = P_a(V_2 - V_1) + \frac{K}{2A^2} [(V_2 - V_0 + V_1 - V_0)(V_2 - V_0 - V_1 + V_0)]$$
$$= P_a(V_2 - V_1) + \frac{K}{2A^2} [(V_2 + V_1 - 2V_0)(V_2 - V_1)]$$

or

$$W = \left[P_a + \frac{K}{2A^2}(V_2 + V_1 - 2V_0) \right] (V_2 - V_1) \quad \text{(B)}$$

Force balance across the piston, in the initial and final states, gives

$$P_1 A = P_a A + \frac{K}{A}(V_1 - V_0)$$

or

$$P_1 = P_a + \frac{K}{A^2}(V_1 - V_0) \quad \text{(C)}$$

Similarly

$$P_2 = P_a + \frac{K}{A^2}(V_2 - V_0) \quad \text{(D)}$$

Adding Eqs. (C) and (D) and dividing by 2, we get

$$\frac{P_1 + P_2}{2} = P_a + \frac{K}{2A^2}(V_1 + V_2 - 2V_0) \quad \text{(E)}$$

Substituting Eq. (E) in Eq. (B), we get

$$W = \frac{(P_1 + P_2)}{2}(V_2 - V_1) = \frac{(200 \times 10^3 + 600 \times 10^3)}{2}(0.2 - 0.1) = 400 \times 10^3 \times 0.1 = 40 \text{ kJ}$$

The pressure versus volume diagram for the gas is shown in Fig. 1.16.

Work done, $W = \int_1^2 PdV$ = area under the curve

$$= \frac{(P_1 + P_2)}{2}(V_2 - V_1) = \text{(area of trapezium A-1-2-B)}$$

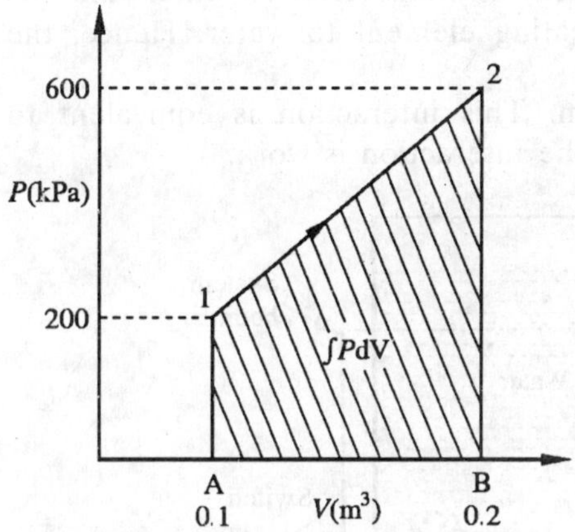

Fig. 1.16 P–V diagram for Example 1.12.

EXAMPLE 1.13 A piston-cylinder assembly contains 1 kg water at 100°C at 101.325 kPa. The piston-cylinder assembly is heated at constant pressure till all the water is converted into steam at 100°C. Given that the specific volumes of liquid water and steam at 100°C are 0.001 04 m^3/kg and 1.673 m^3/kg, respectively, calculate the work done by water.

SOLUTION The liquid water is converted into steam at constant temperature and pressure. The work done is given by

$$W = \int_1^2 PdV = P(V_2 - V_1) = 101.325 \times 10^3(1.673 - 0.001\,04) = 169.41 \text{ kJ}$$

1.16 Heat

Consider a gas contained in a rigid vessel. Suppose the vessel is placed in contact with a hot body and the temperature and pressure of the gas is monitored. We find that the temperature and pressure of the gas increase with time. The system (the gas inside the vessel) has interacted with the surroundings and its energy has changed. Can we call this interaction as work? From a macroscopic point of view, we find that the boundary of the system has not moved and the volume of the gas is constant. Hence there is no work done on the system. Therefore, the interaction between a system and its surroundings which cannot be called work at a macroscopic level is called heat interaction. That is, *energy transfer between a system and its surroundings is called heat if it occurs by virtue of the temperature difference across the boundary.*

The two modes of energy transfer—work and heat—depend on the choice of the system. Heat, like work, is not a property of the system and is a path function.

EXAMPLE 1.14 Suppose an electrical immersion heater is placed in a certain quantity of water contained in a vessel and the heater is switched on to raise the temperature of water. Identify the interaction as work or heat if (a) water alone is considered as the system and (b) the combination of water and the heating element is considered as the system.

SOLUTION (a) The choice of the system is shown in Fig. 1.17. The temperature of water increases after switching on the heater. Therefore, energy is transferred to water and the energy transfer occurs from the hot surface of the heating element to water. Hence, the interaction is heat.

(b) The heating element is also a part of the system. This interaction is equivalent to discharging of a storage battery which is work. Hence, the interaction is work.

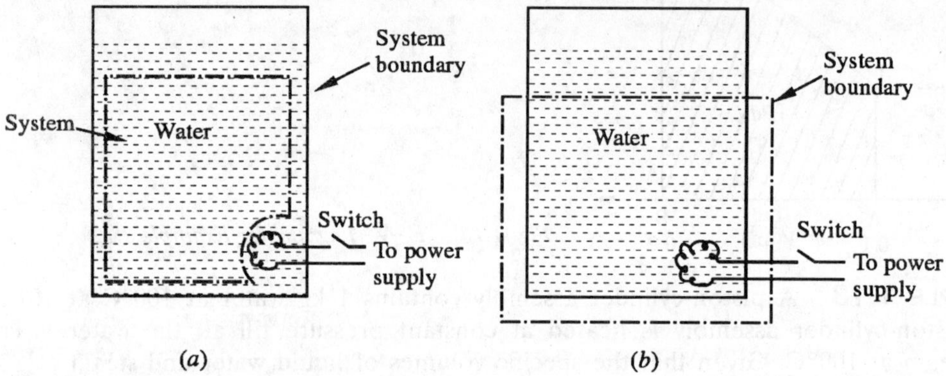

Fig. 1.17 (a) The heating element is excluded from the system and forms part of the surroundings. (b) Water as well as the heating element constitute the system.

The atmospheric pressure and temperature change from place to place and also with the altitude. For convenience, some standard values are needed to specify the properties of certain substances. For this purpose one standard atmosphere is defined as the pressure exerted by a column of 760 mm of mercury at 0°C, where the density of mercury is 13,595 kg/m^3 and the acceleration due to gravity is equal to 9.8067 m/s^2. The standard temperature is chosen as 0°C. This choice is called *standard conditions or normal conditions* and is usually referred as *Standard Temperature and Pressure* (STP) or *Normal Temperature and Pressure* (NTP). At STP, one mole of a gas occupies 22.4 litres.

Review Questions

1. Explain macroscopic and microscopic points of view in the study of thermodynamics.
2. Describe the nature of variables used for describing a thermodynamic system.
3. Define the unit "mole".
4. Liquified Petroleum Gas (LPG) is sold in small cylinders for use as domestic fuel. If one LPG cylinder contains 15 kg gas, express the quantity of gas in the cylinder in moles, treating LPG as pure butane (C_4H_{10}).

5. The rate of heat transfer q is given by the relation $q = kA\Delta T$, where q is in watts and A is in m^2. Determine the SI units of thermal conductivity k.
6. What are the SI units of specific heat?
7. How do you express a millinewton and metre × newton in SI units?
8. If you have to strictly adopt SI units, can you use centimetres to express the length of a pencil? If not, what other units are to be employed?
9. The density of several liquids is usually specified in the units grams per cubic centimetre. Are they in conformity with SI units? If not, specify the correct SI units of density.
10. The van der Waals equation of state is given by $[P + (a/v^2)](v - b) = RT$, where P is pressure, v is specific volume, and the units of the constant R are kJ/kg K. Determine the units of the constants a and b.
11. What are the characteristics of a thermodynamic system?
12. Distinguish between a closed system and an open system.
13. What is meant by an isolated system?
14. Can an adiabatic system exchange energy with its surroundings?
15. What is meant by universe?
16. What conditions should a characteristic satisfy for it to be used as a thermodynamic property?
17. What is meant by a point function?
18. What is a path function?
19. Give a few examples of thermodynamic properties.
20. Give two examples of extensive properties and two examples of intensive properties.
21. Distinguish between extensive and intensive properties.
22. Define a phase.
23. How many phases can be found if liquid water is mixed with kerosene?
24. Suppose a mixture of water and kerosene is boiled in a closed vessel, how many phases can be found?
25. What is meant by equilibrium?
26. Distinguish between steady-state and equilibrium.
27. Suppose one end of a metal rod is immersed in boiling water, while the other end is inserted in melting ice. After a sufficiently long time, the temperature at any specified location in the rod does not undergo any change with time. Is the metal rod in a state of equilibrium? Give reasons.
28. What criterion must be satisfied for a system A to be in thermal equilibrium with system B?
29. What criteria should a system satisfy for it to be in a state of thermodynamic equilibrium?
30. What is a quasi-static process?
31. What is a reversible process?
32. What is the difference between a quasi-static process and a reversible process?
33. How do we classify a process as reversible or irreversible?
34. What is a cyclic process?
35. Name the macroscopic modes of energy and specify why they are so named.
36. What is internal energy?
37. State the thermodynamic definition of work.
38. What is the advantage of the thermodynamic definition over the mechanics definition of work?

39. What is heat interaction?
40. Does work done depend on the path followed by a system?
41. Is heat a property of a system? Can we say that a system contains a certain amount of heat?
42. Which characteristics are common to both work and heat?
43. Do work and heat interactions depend on the choice of a system? Give an example.
44. What is meant by free expansion? What is the work done in the free expansion of a gas?
45. Is the work done always given by $\int PdV$? If not, specify the conditions under which it is equal to $\int PdV$.
46. Give an example where work done is not equal to zero, even if $dV = 0$.
47. Define one standard atmosphere pressure.
48. At atmospheric pressure and 0°C, the density of ice is lower than that of water. If ice melts into water at 0°C, is there any work done? Explain.
49. What is a gauge pressure and how is it related to absolute pressure?
50. How does one refer to pressure below the atmospheric pressure? What is the absolute pressure of a system, if the system is at a vacuum of 70 kPa.
51. What is a continuum?

Problems

1.1 Air can be treated as a mixture of oxygen and nitrogen. The mole fraction of oxygen in air is 0.21. How many moles of air will give a mass of 100 kg?

1.2 Calculate the force needed to accelerate a scooter of mass 85 kg by 3 m/s².

1.3 A domestic refrigerator has a compressor which draws 200 W. On an average, the compressor runs for only 8 hours a day to maintain a constant temperature in the refrigerator. Estimate the energy needed to operate the refrigerator for 30 days.

1.4 A piston-cylinder assembly shown in Fig. 1.18 contains a gas. The diameter of the piston is 10 mm. The pressure exerted by the gas is exactly balanced when a 2 kg mass is placed on the piston. If the atmospheric pressure is 101.325 kPa, calculate the pressure of the gas inside the cylinder.

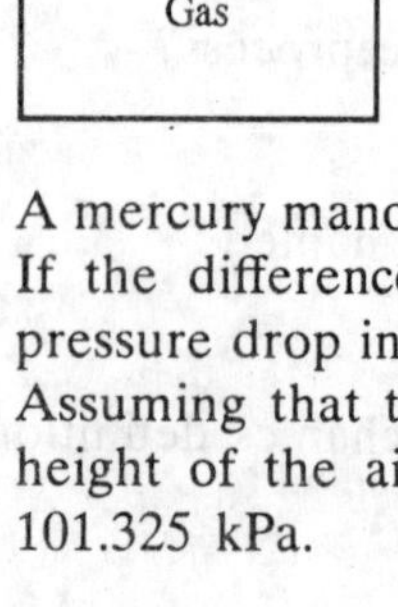

Fig. 1.18 Sketch for Problem 1.4.

1.5 A mercury manometer is used to measure the pressure drop of water flowing in a pipeline. If the difference of the mercury levels in the two limbs reads 0.4 m, determine the pressure drop in SI units. The density of mercury is 13.6×10^3 kg/m³ and $g = 9.81$ m/s².

1.6 Assuming that the atmospheric air has a constant density of 1.16 kg/m³, determine the height of the air column above the earth's surface where the atmospheric pressure is 101.325 kPa.

1.7 A pressure of one standard atmosphere is defined as equal to the pressure exerted by a column of 760 mm mercury (density = 13.5951×10^3 kg/m^3) at a location where g = 9.8067 m/s^2. If water is used instead of mercury in a barometer, determine the height of the water column which gives one standard atmosphere pressure.

1.8 Classify the following as extensive and intensive properties: (a) pressure (b) volume (c) density (d) energy.

1.9 A scooter of mass 85 kg is moving at a speed of 60 km per hour. Calculate the kinetic energy of the scooter.

1.10 If a car of mass 250 kg is moving at a speed of 30 km per hour, calculate the work to be done to increase its speed to 80 km per hour.

1.11 The pressure of a gas in a cylinder is measured with a pressure gauge. If the pressure is found to be 175 kPa gauge pressure, calculate the absolute pressure of the gas.

1.12 A chamber is maintained at vacuum and if a vacuum gauge reads 85 kPa, estimate the absolute pressure in the chamber.

1.13 Express the pressure of 1 mm of mercury column in SI units.

1.14 A projectile is discharged from the tip of a gun at a velocity of 100 m/s. If the projectile is travelling vertically up, determine the maximum height to which it can reach.

1.15 A balloon is connected to a hydrogen cylinder. Initially the balloon is flat. When it is filled with hydrogen, the balloon appears as a sphere of 1 m diameter. If the atmospheric air pressure is 100 kPa, calculate the work done by hydrogen. List the assumptions made, if any.

1.16 One kg of liquid Freon-12 is converted into vapor at the same temperature and pressure. If the specific volumes of liquid and vapor Freon-12 at 0°C and 3.0844 bar are 0.716×10^{-3} m^3/kg and 55.366×10^{-3} m^3/kg, respectively, calculate the work done.

1.17 A particular gas obeys the relation $Pv = RT$, where R = 0.297 kJ/kg K. A piston-cylinder assembly contains 0.1 kg gas at 300 K and 0.1 MPa. If the gas is compressed to a pressure of 1 MPa while holding the temperature constant, determine the work done on the gas.

1.18 One mole of a gas which obeys the relation $Pv = RT$, where R = 8.314 J/mol K, is allowed to expand isothermally at 600 K from 2 MPa to 0.1 MPa, calculate the work done by the gas.

1.19 It is desired to compress one kmol of carbon dioxide which obeys the relation $[P + (a/v^2)](v - b) = RT$ at a constant temperature of 67°C from 28.2 m^3/kmol to a final volume of 7 m^3/kmol. Calculate the work to be done if a = 364.3 kPa (m^3/kmol)^2 and b = 0.0427 m^3/kmol and R = 8.314 kJ/kmol K.

1.20 A gas which is initially at 0.1 MPa and 300 K is to be compressed to a pressure of 1.0 MPa in a compressor. During the compression the pressure and volume are related as $Pv^{1.4}$ = constant. Assume that the gas follows the relation $Pv = RT$, where R = 0.297 kJ/kg K, and estimate the work done per kg of gas.

1.21 A particular gas obeys the relation $Pv = RT$, where v is in m^3/mol and R = 8.314 kJ/kmol K. Initially the gas is at 100 kPa and 300 K. One mole of the gas is heated at constant volume till the pressure is 0.5 MPa and then it is allowed to expand to 0.1 MPa in an adiabatic process. Calculate the net work done by the gas if γ = 1.4.

1.22 A piston-cylinder assembly shown in Fig. 1.15 contains 0.1 m^3 of air at 0.1 MPa. At this stage the spring is just touching the piston and does not exert any force on it. Now, the gas is heated till the volume of the gas is 0.3 m^3 and at this stage, the pressure is

0.9 MPa. As the spring gets compressed, it exerts a force which is proportional to the decrease in its length. Calculate the work done by the gas.

1.23 A particular gas which obeys the relation $Pv = RT$, where v is in m^3/mol and R = 8.314 J/mol K, is heated at constant volume until its temperature is doubled and then cooled at constant pressure till it returns to the original temperature. Finally, the gas is allowed to expand isothermally till it is restored to the original volume. Sketch the path followed by the gas on a pressure versus volume diagram and calculate the net work done by one mole of gas.

1.24 One kg of a gas at 1.0 MPa and 600 K is allowed to expand isothermally to 0.4 MPa and then it is allowed to cool at constant volume to a temperature of 300 K. Calculate the work done by the gas if it obeys the relation $Pv = RT$, with R = 0.297 kJ/kg K.

2 Ideal and Real Gases

2.1 Ideal Gas Equation of State

In chapter 1, we discussed work and heat interactions and it was stated that both work and heat are path functions. If a system undergoes a reversible process, the compression/expansion work done by the system is given by

$$W = \int P dv \tag{2.1}$$

To integrate Pdv, one should know the relation between the pressure and volume of the substance for a particular process under consideration. In this section, we shall study the properties of an ideal gas.

The most convenient method of expressing the behavior of a substance is by a mathematical expression, that is, an equation which represents the P–v–T behavior of a substance. The substances can be present in three different states—namely solids, liquids and gases. The compressibilities of solids and liquids are very small compared to that of gases. Hence, the change in volume of a gas with change in pressure and temperature is quite large. Therefore, first let us study the P–v–T behavior of gases.

Experimental methods were first used in developing an expression of the type $f(P, v, T) = 0$, which is known as the *equation of state*. In 1662, Boyle observed that the volume of a gas is inversely proportional to the pressure at a constant temperature. This is known as *Boyle's law*. That is, at constant temperature

$$v \propto \frac{1}{P} \quad \text{or} \quad Pv = \text{constant} \tag{2.2}$$

In 1787, Charles observed that (a) at constant pressure, the volume of a gas is directly proportional to the temperature and (b) at constant volume, the pressure exerted by a gas is directly proportional to the temperature. This is known as *Charles' law*. That is,

$$\text{At constant pressure:} \quad v \propto T \tag{2.3}$$

$$\text{At constant volume:} \quad P \propto T \tag{2.4}$$

Boyle's law and Charles' law can be combined to yield the relation

$$PV = mRT \quad \text{or} \quad P = \frac{m}{V}RT = \rho RT \tag{2.5}$$

where ρ = density

m = mass of gas

R = specific gas constant, which depends on the gas.

Equation (2.5) is known as the *ideal gas equation of state*. This equation can be rewritten as

$$PV = \left(\frac{m}{M}\right)(MR)T = nR_uT \tag{2.6}$$

where

M = molar mass

R_u = universal gas constant

n = number of moles.

The universal gas constant has a value of 8.314 J/mol K or kJ/kmol K and is related to the specific gas constant by the relation

$$R = \frac{R_u}{M} \tag{2.7}$$

EXAMPLE 2.1 Given that the molar mass of air is 28.97 kg/kmol, calculate the specific gas constant for air.

SOLUTION The specific gas constant R is given by

$$R = \frac{R_u}{M} = \frac{8.314}{28.97} = 0.287 \text{ kJ/kg K}$$

The ideal gas equation of state can be derived from the kinetic theory of gases where the following assumptions are made:

1. The molecules are independent of each other. In other words, there are no attractive forces between the molecules.
2. The molecules do not occupy any volume. That is, the volume occupied by the molecules is quite negligible compared to the volume available for motion of the molecules.

In 1843, Joule demonstrated the fact that the internal energy of an ideal gas is a function of temperature only and is independent of pressure and volume. That is,

$$u = u(T) \tag{2.8}$$

$$\left(\frac{\partial u}{\partial P}\right)_T = 0$$

and

$$\left(\frac{\partial u}{\partial v}\right)_T = 0$$

2.2 Enthalpy and Specific Heat

The enthalpy (H) is defined as

$$H = U + Pv \tag{2.9}$$

For an ideal gas $U = U(T)$ only and $Pv = mRT$ and hence $H = H(T)$ only.

The First Law of Thermodynamics is discussed in chapter 4. For brevity, we state here the First Law of Thermodynamics as

$$dU = dQ - dW \tag{2.10}$$

where dQ is the energy transferred as heat to the system and dW is the work done by the system. The specific heat at constant volume is defined as the amount of energy transferred as heat at constant volume, per unit mass of a system to raise its temperature by one degree. That is,

$$C_v = \left(\frac{dq}{dT}\right)_v \tag{2.11}$$

where C_v = specific heat at constant volume

q = heat interaction per unit mass.

For a constant volume process $dW = 0$ and hence $dq = du$. Therefore, Eq. (2.11) can be rewritten as

$$C_v = \left(\frac{du}{dT}\right)_v \tag{2.12}$$

where u is specific internal energy. The units of specific heat are kJ/kg K. For an ideal gas $u = u(T)$ only and hence

$$du = C_v dT \tag{2.13}$$

The specific heat at constant pressure C_p is defined as the energy transferred as heat at constant pressure, per unit mass of a substance to raise its temperature by one degree. That is,

$$C_p = \left(\frac{dq}{dT}\right)_P \tag{2.14}$$

For a constant pressure process Eq. (2.10) can be rewritten as

$$dQ = dU + dW = dU + Pdv = dU + Pdv + vdP \qquad \text{(since } dP = 0 \text{ at constant pressure)}$$

or $$dQ = dU + d(Pv) = d(U + Pv) = dH \quad \text{or} \quad dq = dh \tag{2.15}$$

From Eqs. (2.14) and (2.15), we can obtain

$$C_p = \left(\frac{\partial h}{\partial T}\right)_P \tag{2.16}$$

Since $h = h(T)$ only for an ideal gas we get

$$dh = C_p dT \tag{2.17}$$

The ratio of specific heats (γ) is given by

$$\gamma = C_p/C_v \tag{2.18}$$

For monatomic ideal gases γ = 1.67 and for diatomic gases γ = 1.4.

2.3 Relation between C_p and C_v

The two specific heats C_p and C_v are related to each other. We know that

$$h = u + Pv \quad \text{or} \quad dh = du + d(Pv)$$

For an ideal gas, the above equation reduces to

$$dh = du + d(RT) = du + RdT \quad \text{or} \quad \frac{dh}{dT} = \frac{du}{dT} + R \quad \text{or} \quad C_p = C_v + R$$

$$\text{or} \quad C_p - C_v = R \quad \text{for an ideal gas} \tag{2.19}$$

We know that

$$\gamma = \frac{C_p}{C_v} = \frac{C_v + R}{C_v} = 1 + \frac{R}{C_v} \quad \text{or} \quad \frac{R}{C_v} = \gamma - 1$$

$$\text{or} \quad C_v = \frac{R}{\gamma - 1} \tag{2.20}$$

$$\text{Then} \quad C_p = \gamma C_v = \frac{R\gamma}{\gamma - 1} \tag{2.21}$$

Therefore, the values of C_p and C_v can be obtained from a knowledge of R and γ.

EXAMPLE 2.2 Determine the specific heats C_p and C_v for the monatomic gas Helium, given that γ = 1.67 and M = 4 kg/kmol for Helium.

SOLUTION We know that

$$R = \frac{R_u}{M} = \frac{8.314}{4} = 2.0785 \text{ kJ/kg K}$$

$$C_v = \frac{R}{\gamma - 1} = \frac{2.0785}{1.67 - 1} = 3.1022 \text{ kJ/kg K}$$

$$C_p = \frac{R\gamma}{\gamma - 1} = \frac{2.0785 \times 1.67}{1.67 - 1} = 5.1807 \text{ kJ/kg K}$$

EXAMPLE 2.3 Calculate the change in internal energy and enthalpy for one kg of Helium when it is changed from 30°C to 230°C, assuming that Helium is an ideal gas.

SOLUTION For Helium (see Example 2.2)

$$C_v = 3.1022 \text{ kJ/kg K} \quad \text{and} \quad C_p = 5.1807 \text{ kJ/kg K}$$

For an ideal gas

$$du = C_v dt \quad \text{or} \quad \Delta u = \int C_v dT = C_v(T_2 - T_1)$$

$$\Delta u = 3.1022 \times 10^3 \times (230 - 30) = 620.44 \text{ kJ/kg}$$

$$dh = C_p dT \quad \text{or} \quad \Delta h = \int C_p dT = C_p(T_2 - T_1)$$

$$\Delta h = 5.1807 \times 10^3 \times (230 - 30) = 1036.14 \text{ kJ/kg}$$

EXAMPLE 2.4 It is desired to store 20 kg ammonia gas at 200°C and 5 MPa in a tank. Assuming that ammonia is an ideal gas, estimate the volume of the tank.

SOLUTION For ammonia $R = 0.4882$ kJ/kg K

$$PV = mRT \quad \text{or} \quad V = \frac{mRT}{P} = \frac{20 \times 0.4882 \times 10^3 \times 473.15}{5 \times 10^6} = 0.924 \text{ m}^3$$

2.4 Real Gases

The ideal gas law is only an approximation to the actual behavior of gases. At high densities, that is, at high pressures and low temperatures, the behavior of actual or real gases deviate from that predicted by the ideal gas law. In general, at sufficiently low pressures or at low densities all gases tend to behave like ideal gases. The assumptions made in deriving the ideal gas law from kinetic theory cannot be justified for a real gas. Several attempts have been made to modify the ideal gas law by incorporating corrections to account for the volume occupied by the molecules as well as to account for the attractive forces between the molecules. One of the earliest attempts to develop an equation of state for real gases was made by van der Waals. The van der Waals equation of state is discussed in the following section.

2.5 van der Waals Equation of State

In 1873, van der Waals proposed an equation of state which took into account the volume occupied by the molecules and the attractive forces between them. The van der Waals equation of state is given by

$$\left(P + \frac{a}{v^2}\right)(v - b) = RT$$

or $$P = \frac{RT}{v - b} - \frac{a}{v^2} \tag{2.22}$$

where a and b are van der Waals constants. The van der Waals equation of state is cubic in volume and in general there will be three values of v for given values of P and T. However, in some range of values of P and T there is only one real value of v. For $T > T_c$ (critical temperature) there will be only one real value of v and for $T < T_c$ there will be three real values. For a van der Waals gas, the isotherms are shown in Fig. 2.1.

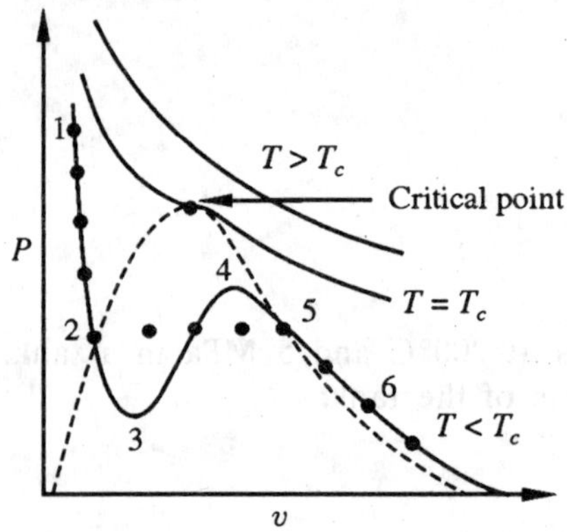

Fig. 2.1 *P–v diagram showing the isotherms for a van der Waals gas.*

In Fig. 2.1, the solid curves represent the values predicted by the van der Waals equation of state and the points represent the experimentally determined values. It can be observed from Fig. 2.1 that at temperatures greater than the critical temperature, there is only one real value of volume for given values of P and T. However, at temperatures less than the critical temperature (for example the isotherm 123456) there are three real values of volume for a given value of P and T. The experimental values differ from those predicted by van der Waals equation of state in the region 2345 if $T < T_c$. It can be shown (the proof is beyond the scope of this book) that the region 2345 is not physically realizable. The physically acceptable isotherm is 1256, where the states 2 and 5 are connected by a horizontal line.

The van der Waals constants a and b can be determined from experimental P–v–T data. At constant volume, Eq. (2.22) represents a straight line with slope $[R/(v - b)]$ and intercept $(-a/v^2)$, where P is plotted against T as shown in Fig. 2.2. From a knowledge of the slope and the intercept, one can determine the values of a and b.

Alternatively, one can use the criterion that the critical isotherm (isotherm passing through the critical point) shows a point of inflexion. Stated mathematically

$$\left(\frac{\partial P}{\partial v}\right)_{T=T_c} = 0 \quad \text{and} \quad \left(\frac{\partial^2 P}{\partial v^2}\right)_{T=T_c} = 0 \tag{2.23}$$

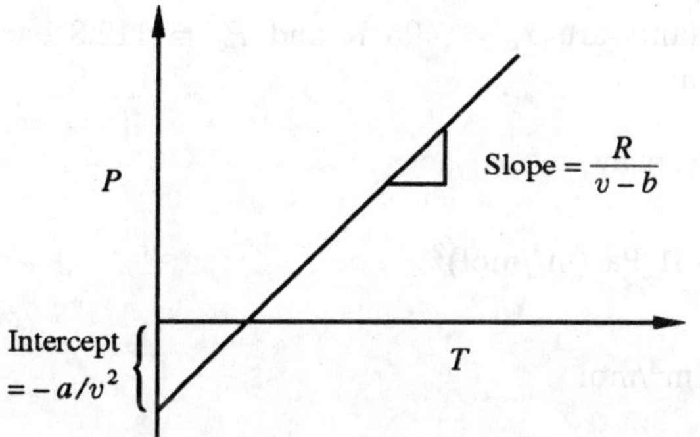

Fig. 2.2 *P versus 1/T diagram for van der Waals gas at constant volume.*

$$\left(\frac{\partial P}{\partial v}\right)_{T=T_c} = \frac{-RT_c}{(v_c - b)^2} + \frac{2a}{v_c^3} = 0$$

or

$$\frac{RT_c}{(v_c - b)^2} = \frac{2a}{v_c^3} \tag{2.24}$$

$$\left(\frac{\partial^2 P}{\partial v^2}\right)_{T=T_c} = \frac{2RT_c}{(v_c - b)^3} - \frac{6a}{v_c^4} = 0$$

or

$$\frac{2RT_c}{(v_c - b)^3} = \frac{6a}{v_c^4} \tag{2.25}$$

Dividing Eq. (2.25) with Eq. (2.24), we get

$$\frac{2}{v_c - b} = \frac{3}{v_c} \quad \text{or} \quad v_c = 3b \tag{2.26}$$

At the critical point, the van der Waals equation is given by

$$P_c = \frac{RT_c}{v_c - b} - \frac{a}{v_c^2} \tag{2.27}$$

Substituting the value of v_c from Eq. (2.26) in Eqs. (2.24) and (2.27), we get

$$T_c = \frac{8a}{27Rb} \quad \text{and} \quad P_c = \frac{a}{27b^2} \tag{2.28}$$

From Eqs. (2.26) and (2.28), we can obtain

$$a = \frac{27R^2T_c^2}{64P_c} \quad \text{and} \quad b = \frac{RT_c}{8P_c} \tag{2.29}$$

Thus, one can determine the constants a and b from a knowledge of the critical constants. The critical constants of some selected substances are presented in Appendix (Table A1)

EXAMPLE 2.5 For ammonia, the critical constants are T_c = 406 K and P_c = 112.8 bar. Calculate the van der Waals constants for ammonia.

SOLUTION The van der Waals constants are given by

$$a = \frac{27R^2T_c^2}{64P_c} = \frac{27 \times (8.314)^2 \times (406)^2}{64 \times 112.8 \times 10^5} = 0.4261 \text{ Pa (m}^3\text{/mol)}^2$$

$$b = \frac{RT_c}{8P_c} = \frac{8.314 \times 406}{8 \times 112.8 \times 10^5} = 37.406 \times 10^{-6} \text{ m}^3\text{/mol}$$

EXAMPLE 2.6 The van der Waals constants for ethane are a = 0.557 Pa (m^3/mol)^2 and b = 64.99 × 10^{-6} m^3/mol. Estimate the critical temperature and critical pressure for ethane.

SOLUTION We know that (see Eq. (2.28))

$$T_c = \frac{8a}{27Rb} \quad \text{and} \quad P_c = \frac{a}{27b^2}$$

Therefore $$T_c = \frac{8 \times 0.557}{27 \times 8.314 \times 64.99 \times 10^{-6}} = 305.4 \text{ K}$$

$$P_c = \frac{0.557}{27 \times (64.99 \times 10^{-6})^2} = 48.84 \text{ bar}$$

The estimation of pressure or temperature, if the volume is given, from the van der Waals equation is straightforward. Since the van der Waals equation is cubic in volume, the estimation of volume from the knowledge of pressure and temperature requires an iterative calculation. To facilitate the iterative calculations, the van der Waals equation can be rearranged as

$$v = \frac{RT}{P + a/v^2} + b$$

For iterative calculations, the above equation can be expressed as

$$v_{i+1} = \frac{RT}{P + a/v_i^2} + b \tag{2.30}$$

where v_i = volume in the ith iteration

v_{i+1} = volume in the $(i + 1)$th iteration.

To initiate the iterative calculation, when i = 0, v_0 can be taken as equal to that predicted by the ideal gas law. That is

$$v_0 = \frac{RT}{P}$$

Then calculate v_1 from Eq. (2.30). Compare v_1 with v_0. If they are approximately equal, then the volume is given by v_1 or else repeat the calculations till $v_{i+1} = v_i$. The method of calculation is illustrated in the following example.

EXAMPLE 2.7 A cylinder of volume 0.1 m^3 is filled with ethane at 350 K and 50 bar pressure. Estimate the mass of ethane in the cylinder, using the van der Waals equation of state.

SOLUTION The van der Waals constants for ethane are a = 0.5562 Pa(m^3/mol)^2 and $b = 64 \times 10^{-6}$ m^3/mol. First let us calculate the volume of 1 mol ethane.

$$v_0 = \frac{RT}{P} = \frac{8.314 \times 350}{50 \times 10^5} = 5.82 \times 10^{-4} \text{ m}^3\text{/mol}$$

$$v_{i+1} = \frac{RT}{(P + a/v_i^2)} + b$$

For the first iteration,

$$v_1 = \frac{RT}{(P + a/v_0^2)} + b = \frac{8.314 \times 350}{50 \times 10^5 + \dfrac{0.5562}{(5.82 \times 10^{-4})^2}} + 64 \times 10^{-6} = 5.02 \times 10^{-4} \text{ m}^3\text{/mol}$$

$v_1 \neq v_0$. Hence repeat the calculations.

$$v_2 = \frac{RT}{(P + a/v_1^2)} + b = \frac{8.314 \times 350}{50 \times 10^5 + \dfrac{0.5562}{(5.02 \times 10^{-4})^2}} + 64 \times 10^{-6} = 4.68 \times 10^{-4} \text{ m}^3\text{/mol}$$

Since $v_2 \neq v_1$, the iterative calculations are to be continued. The results of calculations are shown below.

Iteration	0	1	2	3	4	5	6	7
Volume (m^3/mol)$\times 10^4$	5.82	5.02	4.68	4.5	4.4	4.34	4.3	4.3

The calculated values of v in the 6th and 7th iterations agree with each other.

Hence $v = 4.3 \times 10^{-4}$ m^3/mol

$$\text{Quantity of ethane in cylinder} = \frac{0.1}{4.3 \times 10^{-4}} = 232.56 \text{ mol}$$

Molar mass of ethane = 30.07 kg/kmol

Mass of ethane = $232.56 \times 10^{-3} \times 30.07$ = 6.99 kg

2.6 Compressibility Factor

The deviation from ideal behavior of a gas is expressed in terms of the compressibility factor

Z, which is defined as the ratio of the actual volume to the volume predicted by the ideal gas law. That is

$$Z = \frac{\text{Actual volume}}{\text{volume predicted by ideal gas law}} = \frac{v}{RT/P} = \frac{Pv}{RT} \tag{2.31}$$

For an ideal gas $Pv = RT$ and hence $Z = 1$ at all temperatures and pressures. The experimental P–v–T data is used to prepare the compressibility chart. In the compressibility chart, the compressibility factor Z is plotted as a function of reduced pressure for various values of reduced temperature. The reduced parameters are defined as

$$\text{Reduced pressure, } P_R = \frac{P}{P_c} \tag{2.32}$$

$$\text{Reduced temperature, } T_R = \frac{T}{T_c} \tag{2.33}$$

$$\text{Reduced volume, } v_R = \frac{v}{v_c} \tag{2.34}$$

where P_c, T_c and v_c denote the critical pressure, critical temperature and critical volume respectively. In the preparation of the compressibility chart, it was found to be convenient to use pseudo-reduced volume instead of reduced volume. The pseudo-reduced volume v'_R is defined as

$$\text{pseudo-reduced volume, } v'_R = \frac{v}{(RT_c/P_c)} = \frac{P_c v}{RT_c} \tag{2.35}$$

The compressibility chart is presented in Appendix (Figs. A1–A2). The use of the compressibility chart is illustrated in the following examples.

EXAMPLE 2.8 Rework Example 2.7 using the compressibility chart.

SOLUTION For ethane, $T_c = 305.4$ K and $P_c = 48.8$ bar

$$P_R = \frac{P}{P_c} = \frac{50}{48.8} = 1.02$$

$$T_R = \frac{T}{T_c} = \frac{350}{305.4} = 1.15$$

Read the value $Z = 0.74$ from the compressibility chart at $P_R = 1.02$ and $T_R = 1.15$ as shown in Fig. 2.3.

$$Z = \frac{Pv}{RT} \quad \text{or} \quad v = \frac{ZRT}{P}$$

or $$v = \frac{0.74 \times 8.314 \times 10^3 \times 350}{50 \times 10^5} = 0.4307 \text{ m}^3\text{/kmol}$$

$$\text{Quantity of ethane} = \frac{0.1}{0.4307} = 0.2322 \text{ kmol}$$

$$\text{Mass of ethane} = 0.2322 \times 30.07 = 6.98 \text{ kg}$$

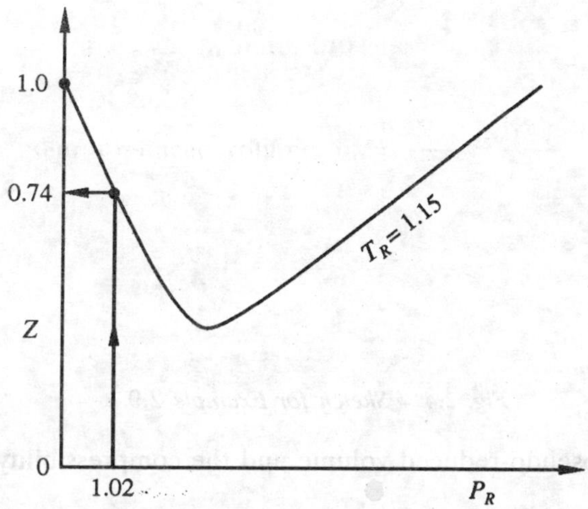

Fig. 2.3 *Sketch for Example 2.8.*

EXAMPLE 2.9 If a cylinder of 0.1 m³ volume is to be filled with 7 kg ethane at 350 K, determine the pressure to which the cylinder is to be charged. Use the compressibility chart to solve the problem.

SOLUTION For ethane, T_c = 305.4 K and P_c = 48.8 bar

$$T_R = \frac{T}{T_c} = \frac{350}{305.4} = 1.15$$

$$\text{Quantity of ethane} = \frac{7}{30.07} = 0.2328 \text{ kmol}$$

To read the value of Z, we need to know P_R and T_R. However, P_R cannot be calculated since P is the quantity to be estimated. To circumvent this difficulty, we can proceed as follows:

$$P_R = \frac{P}{P_c} \quad \text{or} \quad P = P_R P_c = 48.8 \times 10^5 P_R$$

$$Z = \frac{Pv}{RT} = \frac{PV}{nRT} = \frac{48.8 \times 10^5 P_R V}{nRT} = \frac{48.8 \times 10^5 \times 0.1 P_R}{0.2328 \times 10^3 \times 8.314 \times 350}$$

or $$Z = 0.72 P_R \qquad \text{(A)}$$

Equation (A) represents a straight line passing through the origin with a slope of 0.72. Therefore, draw a straight line of slope 0.72 passing through the origin on the compressibility chart as shown in Fig. 2.4 and locate the point of intersection of this line with $T_R = 1.15$ line and read $P_R = 1.1$. Then $P = 48.8 \times 10^5\ P_R = 48.8 \times 10^5 \times 1.1 = 53.68$ bar.

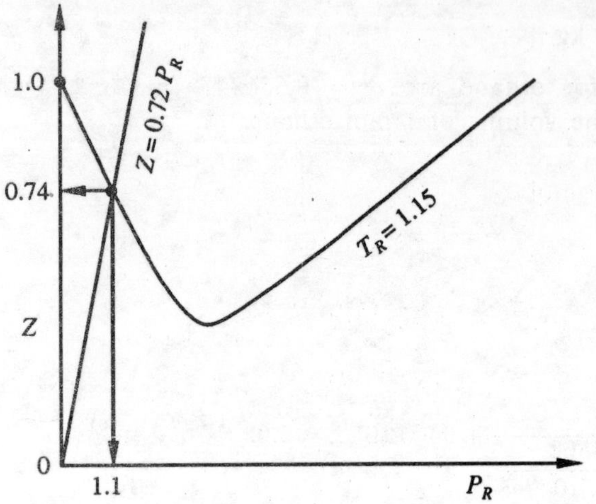

Fig. 2.4 Sketch for Example 2.9.

EXAMPLE 2.10 Rework Example 2.9 using pseudo-reduced volume and the compressibility chart.

SOLUTION $T_R = 1.15$ (see Example 2.9). Pseudo-reduced volume is given by

$$v'_R = \frac{P_c v}{RT_c} = \frac{P_c(V/n)}{RT_c} = \frac{48.8 \times 10^5 (0.1/0.2328 \times 10^3)}{8.314 \times 305.4} \quad \text{or} \quad v'_R = 0.83$$

Then read $P_R = 1.02$ at the point of intersection of $T_R = 1.15$ curve and $v'_R = 0.83$ curve as shown in Fig. 2.5.

$$P = 48.8 \times 10^5 P_R = 48.8 \times 10^5 \times 1.02 = 49.78 \text{ bar}$$

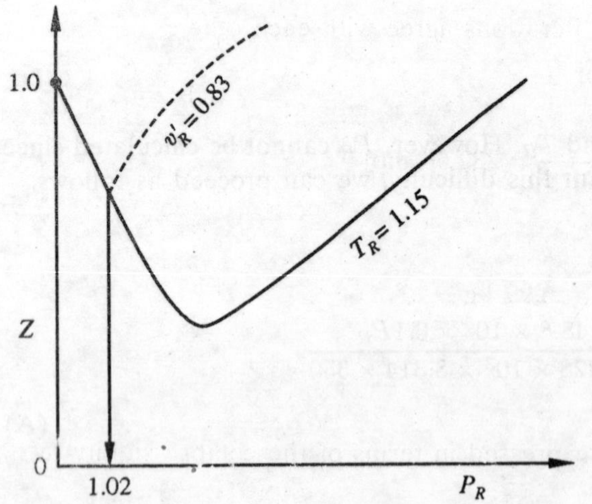

Fig. 2.5 Sketch for Example 2.10.

EXAMPLE 2.11 If a cylinder of volume 0.1 m^3 contains 7 kg ethane at a pressure of 50 bar, estimate the temperature of ethane in the cylinder, using the compressibility chart.

SOLUTION For ethane, $T_c = 305.4$ K and $P_c = 48.8$ bar; $M = 30.07$ kg/kmol

Quantity of ethane, $n = \dfrac{m}{M} = \dfrac{7}{30.07} = 0.2328$ kmol

$$P_R = \frac{P}{P_c} = \frac{50}{48.8} = 1.02$$

Pseudo-reduced volume, $v'_R = \dfrac{P_c v}{RT_c} = \dfrac{P_c(V/n)}{RT_c} = \dfrac{48.8 \times 10^5(0.1/0.2328 \times 10^3)}{8.314 \times 305.4} = 0.83$

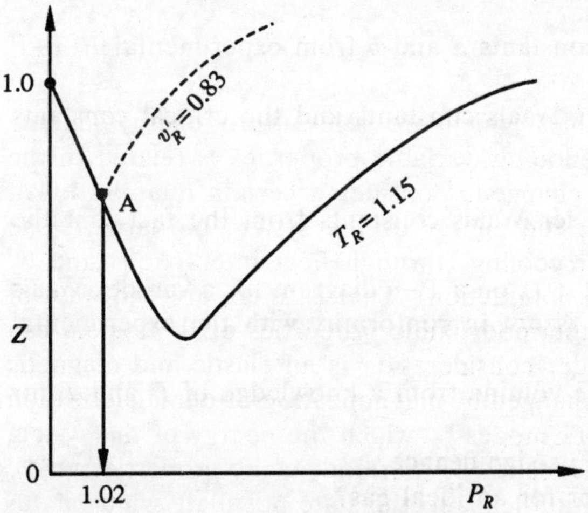

Fig. 2.6 Sketch for Example 2.11.

Draw a vertical line at $P_R = 1.02$ on the compressibility chart as shown in Fig. 2.6 and locate the point of intersection of this line with $v'_R = 0.83$ curve (shown as point A in Fig. 2.6). The curve $T_R = 1.15$ passes through this point A.

$$T_R = 1.15 = T/T_c = T/305.4 \quad \text{or} \quad T = 351.2 \text{ K}$$

Review Questions

1. What is the most convenient method of expressing the P–v–T behavior of a pure substance?
2. What is Boyle's law?
3. What is Charles' law?
4. What is an equation of state?
5. What is the ideal gas equation of state?
6. What is the universal gas constant and how does it differ from the specific gas constant?
7. What assumptions are involved in deriving the ideal gas equation of state from the kinetic theory?

8. Define enthalpy?
9. Define the specific heat at constant volume and the specific heat at constant pressure, in terms of properties u and h.
10. What is the ratio of specific heats (γ) for a monatomic gas and for a diatomic gas?
11. What is the difference between C_p and C_v for an ideal gas?
12. Does the internal energy of an ideal gas depend on pressure?
13. Is it correct to state that the enthalpy of an ideal gas depends on temperature and pressure? If not, on what factors does the enthalpy of an ideal gas depend?
14. Is the specific heat of an ideal gas independent of temperature?
15. What is a real gas and how does it differ from an ideal gas?
16. Which factors were taken into account by van der Waal in modifying the equation of state?
17. What is van der Waals equation of state?
18. How many values for volume can be obtained, for a given pressure and temperature, from the van der Waals equation of state? Explain.
19. How can one determine the van der Waals constants a and b from experimental P–v–T data?
20. What is the relationship between the van der Waals constants and the critical constants of a gas?
21. What is an inflexion point?
22. Explain the method of determining the van der Waals constants from the fact that the critical isotherm exhibits a point of inflexion.
23. Draw the isotherms for $T > T_c$, $T = T_c$ and $T < T_c$ on a P–v diagram for a van der Waals gas and explain why the isotherm at $T < T_c$ is not in conformity with the experimental observations.
24. Develop an iterative method of estimating the volume from a knowledge of P and T for a van der Waals gas.
25. Define the compressibility factor and explain its significance.
26. What is the value of the compressibility factor for an ideal gas?
27. Can a real gas have a compressibility factor equal to unity?
28. Define reduced parameters.
29. What is pseudo-reduced volume and why is it introduced?
30. Compare the value of the pressure exerted by a real gas with that of an ideal gas if $Z < 1$ and explain the reason for such behavior of a real gas.
31. Compare the value of volume occupied by a real gas with that of an ideal gas if $Z > 1$ and explain the reason for such a behavior of a real gas.
32. What is the value of the compressibility factor at the critical point for a van der Waals gas? Does it agree with the value reported in the compressibility chart?
33. Explain the method of estimating the pressure from known values of temperature and volume for a real gas, if the compressibility factor is not known.
34. Sketch the compressibility chart.

Problems

2.1 Using the ideal gas law estimate the densities of Helium, air and carbon dioxide at 0.1 MPa and 300 K.

2.2 Given that the universal gas constant is 8.314 kJ/kmol K, determine the specific gas constants for Helium, air and carbon dioxide.

2.3 Given that γ is 1.667 for a monatomic gas and 1.4 for a diatomic gas, estimate the values of C_p and C_v for Helium and air.

2.4 Derive the relation $C_p - C_v = R$ for an ideal gas.

2.5 Estimate the change in internal energy and enthalpy of 10 kg air if air is changed from 25°C to 525°C.

2.6 50 kg methane is stored in a vessel of volume 0.5 m^3 at 25°C. Determine the pressure developed by the gas if (a) methane is an ideal gas. (b) methane obeys the van der Waals equation of state.

2.7 Estimate the van der Waals constants for carbon dioxide and propane from the critical constants.

2.8 If the van der Waals constants for methane are a = 0.2283 Pa (m^3/mol)^2 and b = 4.3×10^{-5} m^3/mol, estimate the critical constants for methane.

2.9 Assuming that carbon dioxide is a van der Waals gas, estimate the specific volume of CO_2 at 0.5 MPa and 100°C.

2.10 It is required to store 10 kg ethylene at 27°C and 10 bar. Assuming that ethylene is a van der Waals gas, determine the volume of the tank which can hold the gas.

2.11 Determine the compressibility factor for n-octane at 5 MPa and 400°C.

2.12 100 kg of CO_2 is to be stored in a tank at 100°C and 10 MPa. Using the compressibility chart, estimate the size of the tank.

2.13 A tank of volume 1 m^3 is to be filled with propane at 6 MPa and 150°C. Make use of the compressibility chart and estimate the mass of propane that can be charged into the tank.

2.14 Freon-12 for use in domestic refrigerators is usually sold in small cylinders of 3 litres volume. If the cylinder is charged at 10 MPa and 450 K, estimate the mass of Freon-12 (CCl_2F_2). Make use of the compressibility chart.

2.15 The molar masses of nitrogen and ethylene are the same. A cylinder which can hold 7 kg nitrogen at 8 MPa and 300 K is filled with ethylene at 8 MPa and 300 K. Make use of the compressibility chart and estimate the mass of ethylene in the cylinder.

2.16 A rigid vessel of 0.1 m^3 capacity is filled with ethane at 10 MPa and 367 K. If the vessel is cooled to 320 K, determine the pressure of the gas in the cylinder. Use the compressibility chart.

2.17 A rigid vessel of volume 0.1 m^3 is filled with ethylene at 10 MPa and 400 K. If the vessel is cooled to 300 K, estimate the pressure of ethylene in the tank using the compressibility chart.

2.18 A steel vessel of 0.5 m^3 capacity is filled with propane at 85 bar and 518 K. When the vessel is cooled, the gas pressure has reduced to 64 bar. Making use of the compressibility chart, estimate the final temperature of propane in the tank.

2.19 A gas cylinder is filled with ethylene at 340 K and 102.4 bar. The cylinder is left in the open air and after sometime the gas pressure is observed to be 64 bar. Making use of the compressibility chart, estimate the final temperature of ethylene in the cylinder.

3 ZEROTH LAW OF THERMODYNAMICS

A thermodynamic system is defined as a fixed quantity of matter on which we focus our attention. How does one specify the system without any ambiguity? For this purpose one has to describe the system in terms of certain properties. We have noticed that an ideal gas is described by the relation $Pv = RT$. That is, one can independently choose any two of the variables P, v and T to describe the state of an ideal gas. Now, the question that arises is that for a given thermodynamic system, how many properties can be varied independently. This is specified by the state postulate. In this chapter we shall discuss the state postulate, the Zeroth Law of Thermodynamics and temperature scale.

3.1 State Postulate

The thermodynamic properties are those variables which are relevant to energy and its transformation. Therefore, the number of independently variable properties is related to the number of ways the energy of a system can be changed. Consider a certain quantity of an ideal gas as the system. In how many ways can the energy of the system be varied? The energy of the gas can be changed by heating or cooling (through heat interaction) and by performing compression or expansion work (work interaction). That is, there are two ways of changing the energy independently. Hence, two thermodynamic properties of the system can be independently varied. Suppose the system under consideration is an elastic and magnetic substance. Then the energy of the system can be changed by magnetization or demagnetization and also by stretching. That is, there are two work modes by which the energy of the system can be changed in addition to heat interaction. Hence, three thermodynamic properties of the system can be independently varied. This observation leads to the state postulate which states that for a given thermodynamic system, the number of independently variable thermodynamic properties to be specified for specifying the state of a system is equal to the number of work modes plus one. Thus, the state of the system is completely described by specifying that many thermodynamic properties which can be independently varied.

3.2 Zeroth Law of Thermodynamics

The Zeroth Law of Thermodynamics states that if each of the two given systems A and B are in thermal equilibrium with a third system C, then the two systems A and B are in thermal equilibrium with each other.

For a logical presentation of the laws of thermodynamics, the above law has to be dealt with first before discussing the other laws. The First Law of Thermodynamics was proposed by Joule in 1851 followed by the Second Law of Thermodynamics. The Zeroth Law of Thermodynamics was formulated by Fowler in 1931, much after the first and second laws of thermodynamics were proposed and were in use. Therefore, to avoid any confusion the law proposed by Fowler was named as the Zeroth Law of Thermodynamics.

3.3 Temperature Scale

The Zeroth Law of Thermodynamics provides a basis for temperature measurement. The temperature of a group of bodies can be compared by bringing a particular body, (say, a thermometer) into contact with each of them in turn. When two bodies are in thermal equilibrium with each other, their temperatures are equal. To quantify the measurement, the instrument should have an easily observable and measurable characteristic called *thermometric property*. The widely used thermometric properties are—length of a mercury column in a capillary tube, the resistance of a wire, the emf generated at a junction of two dissimilar metal wires, the pressure of a gas enclosed in a constant volume bulb, etc. To assign numerical values to the temperature, it is necessary to establish a temperature scale. The temperature scale is established by assigning numerical values to certain easily reproducible states called fixed points. The commonly used fixed points are:

(a) *Ice point*—the equilibrium temperature of ice with air saturated water at one standard atmosphere pressure, which is assigned a value of 0°C.
(b) *Steam point*—the equilibrium temperature of pure water with its own vapor at one standard atmosphere pressure, which is assigned a value of 100°C.

The temperature scale which assigns 0°C to the ice point and 100°C to the steam point and divides the temperature interval between these two into 100 equal parts is called the *Celsius temperature scale*, named after Anders Celsius (1701–1744) who devised it. This scale was previously called the centigrade scale.

The temperature scale which assigns 32°F to the ice point and 212°F to the steam point and divides the temperature interval between these two into 180 equal parts is called the *Fahrenheit temperature scale*, named after Daniel Gabriel Fahrenheit (1668–1736), who devised it.

In the Celsius temperature scale, the magnitude of one degree is equal to the magnitude of one degree in the Kelvin scale. In the Fahrenheit temperature scale, the magnitude of one degree is equal to the magnitude of one degree in the Rankine scale.

The relationships between the Celsius, Fahrenheit, Kelvin and Rankine scales are shown in Fig. 3.1.

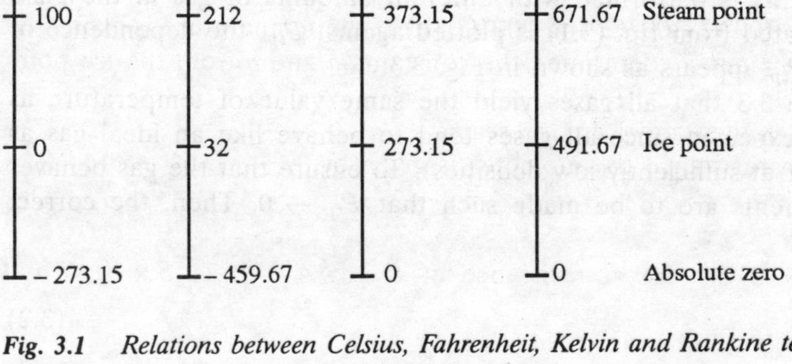

Fig. 3.1 Relations between Celsius, Fahrenheit, Kelvin and Rankine temperature scales.

$$K = 273.15 + °C \quad R = 459.67 + °F \quad °F = 1.8 \times °C + 32$$

where K = temperature in Kelvin scale, R = temperature in Rankine scale.

3.4 Perfect Gas Scale

All thermometers—mercury-in-glass, alcohol-in-glass, thermocouples, resistance thermometers, etc., indicate the same temperature at 0°C and similarly these thermometers indicate the same value at 100°C, because ice point and steam point are used in calibrating them. However, these thermometers indicate different values at any other temperature. This is so because different materials have different temperature–volume relationship or different temperature–emf relationship or different temperature–resistance relationship, and the variation of the thermometric property with temperature may not be linear. Therefore, it is necessary to devise a temperature scale which is independent of the thermometric property of the material.

The perfect gas temperature scale is devised to overcome this drawback. The perfect gas temperature scale is based on the fact that the temperature of an ideal gas held at constant volume is directly proportional to its pressure. A constant volume gas thermometer shown in Fig. 3.2 can be used to measure the temperature of a body. Helium or some other gas is enclosed in the glass bulb B. The bulb B is placed in the system, the temperature of which is to be measured. Then the mercury column is adjusted either by raising or lowering the limb C, which is connected to the glass bulb B through a flexible tubing, such that the mercury level is at the reference mark S. This ensures that the volume of the gas is constant. Then the pressure P exerted by the gas is measured. Similar measurement is carried out by placing the bulb B in a bath which is maintained at triple point of water. Let the pressure of the gas at these conditions be denoted by P_{tp}. The temperature of the triple point of water has been assigned a value of 273.16 K. The ideal gas thermometer is calibrated using only a single fixed point, which is the triple point of water. The temperature T of the system is then given by

$$\frac{T}{T_{tp}} = \frac{P}{P_{tp}}$$

$$\text{or} \quad T = 273.16\frac{P}{P_{tp}} \tag{3.1}$$

where T_{tp} is the triple point temperature of water.

Suppose a series of experiments is performed with different amounts of gas in the glass bulb and the temperature calculated from Eq. (3.1) is plotted against P_{tp}, the dependence of the calculated temperature on P_{tp} appears as shown in Fig. 3.3.

It can be observed from Fig. 3.3 that all gases yield the same value of temperature as $P_{tp} \to 0$. This behavior can be expected since all gases tend to behave like an ideal gas as the pressure approaches zero (or at sufficiently low densities). To ensure that the gas behaves like an ideal gas, the measurements are to be made such that $P_{tp} \to 0$. Then, the correct temperature is given by

$$T = 273.16 \lim_{P_{tp} \to 0} \left(\frac{P}{P_{tp}}\right) \tag{3.2}$$

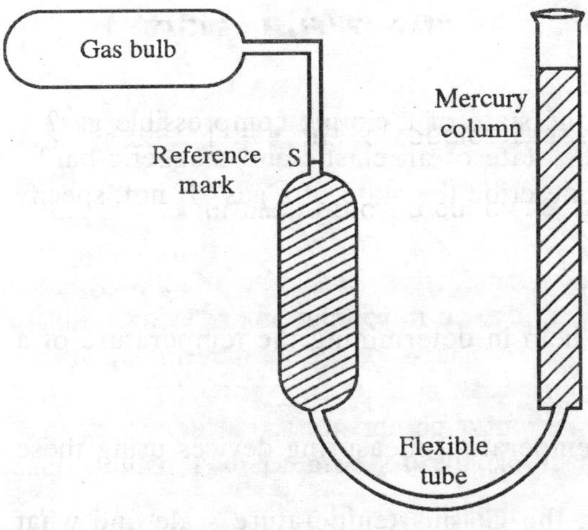

Fig. 3.2 Schematic diagram of a constant volume gas thermometer.

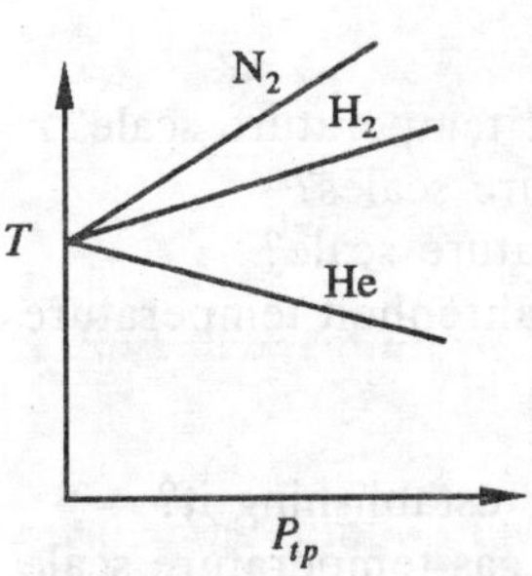

Fig. 3.3 The dependence of calculated temperature on P_{tp}.

3.5 International Practical Temperature Scale (IPTS)

Great skill and expertise is needed to use the ideal gas thermometer. Moreover many corrections are to be applied to obtain precise and accurate results. For practical temperature measurement a more convenient temperature scale is needed. The International Committee on Weights and Measures adopted the International Practical Temperature Scale in 1968 (IPTS-68) based on a number of easily reproducible fixed points. Some of the fixed points on which IPTS-68 is based are presented in Table 3.1.

Table 3.1 Some of the fixed points for IPTS-68.

Fixed point	T(°C)
Triple point of hydrogen	– 259.340
Normal boiling point of hydrogen	– 252.870
Triple point of oxygen	– 218.789
Normal boiling point of oxygen (oxygen point)	– 182.962
Triple point of water	0.010
Normal freezing point of zinc	419.580
Normal freezing point of silver (silver point)	961.930
Normal freezing point of gold (gold point)	1064.430

Review Questions

1. What is the state postulate?
2. How many properties are required to describe the state of a simple compressible gas?
3. How many properties are needed to describe the state of an elastic and magnetic bar?
4. Can we choose ρ and v as the two properties to describe the state of a gas? If not, specify why?
5. State the Zeroth Law of Thermodynamics.
6. What is thermal equilibrium?
7. How does the Zeroth Law of Thermodynamics help in determining the temperature of a body?
8. What is a thermometric property?
9. Name a few thermometric properties and the temperature-measuring devices using these thermometric properties.
10. How many fixed points are used in establishing the Celsius temperature scale and what are they?
11. What is ice point and what is steam point?
12. How does the Celsius temperature scale differ from the Fahrenheit temperature scale?
13. What is the relation between the Celsius and Fahrenheit temperature scales?
14. How is the Kelvin temperature scale related to the Celsius temperature scale?
15. What is the Rankine temperature scale and how is it related to the Fahrenheit temperature scale?
16. What is the relation between degree kelvin and degree rankine?
17. What is the perfect gas temperature scale and what is the basis for establishing it?
18. How many fixed points are used in the establishment of the perfect gas temperature scale and what are they?
19. Describe the procedure for measuring the temperature of a body with the help of a constant volume gas thermometer.
20. What is IPTS-68?

Problems

3.1 One junction of a thermocouple is kept at the ice point and the other junction is at t°C. The emf (ε) expressed in millivolts of the thermocouple is given by

$$\varepsilon = 0.20t - 5.0 \times 10^{-4}t^2$$

(a) Calculate the emf if $t = -100$°C, 100°C, and 200°C. (b) If the temperature is expressed as a linear function of ε as $t = a\varepsilon + b$, determine the constants a and b from the ice point and steam point.

3.2 Express 72°C in Fahrenheit, Kelvin and Rankine scales.

3.3 The resistance of a platinum resistance thermometer is given by

$$R_t = R_0(1 + At + Bt^2)$$

where R_0 = resistance at 0°C, R_t = resistance at t°C, $A = 5.1455 \times 10^{-3}$ and $B = -1.455 \times 10^{-6}$.

Determine the value R_t/R_0 at sulfur point (444.6°C).

4 First Law of Thermodynamics

4.1 First Law of Thermodynamics

A series of experiments carried out by Joule between 1843 and 1848 form the basis for the First Law of Thermodynamics. Of the several experiments conducted by Joule, the paddle wheel experiment is considered as the classic. In the paddle wheel experiment, a certain quantity of liquid at a known temperature is taken in a rigid and insulated container. A paddle wheel is rotated inside the liquid as shown in Fig. 4.1(a). The amount of work done on the liquid is estimated by the distance through which the mass is dropped and the final temperature of the liquid is noted. Then the container's insulation is removed and it is brought into thermal contact with a water bath, till the liquid in the container attains its initial temperature (see Fig. 4.1(b)). The quantity of energy transferred from the liquid to the water bath is estimated by measuring the temperature rise and the mass of water bath.

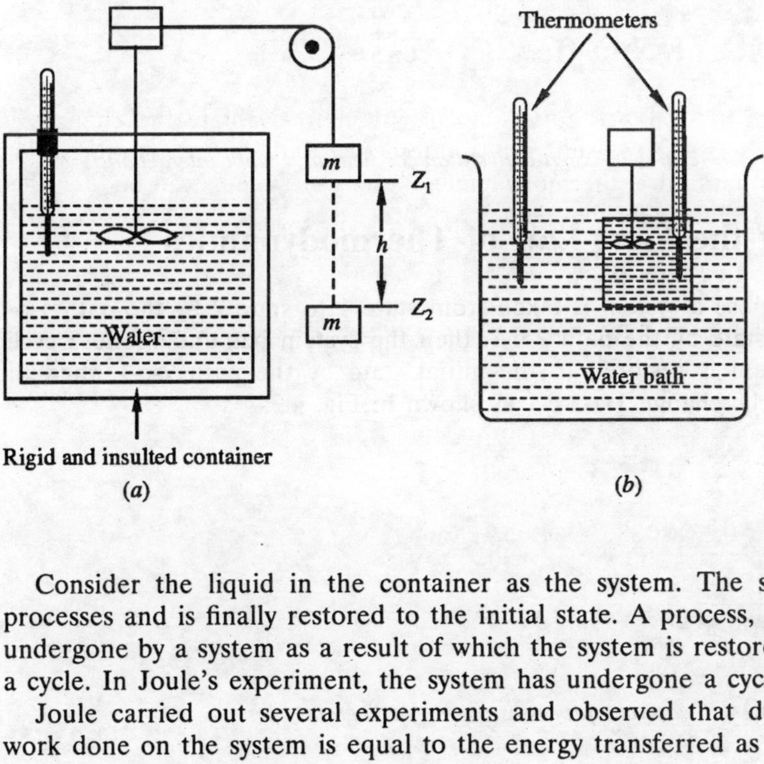

Fig. 4.1 Schematic arrangement of Joule's paddle wheel experiment.

Consider the liquid in the container as the system. The system has undergone certain processes and is finally restored to the initial state. A process, or a combination of processes undergone by a system as a result of which the system is restored to the initial state, is called a cycle. In Joule's experiment, the system has undergone a cyclic change.

Joule carried out several experiments and observed that during a cyclic change the net work done on the system is equal to the energy transferred as heat from the system. That is

$$\oint dQ_{\text{from system}} = \oint dW_{\text{on system}}$$

or
$$\oint dQ_{\text{from system}} - \oint dW_{\text{on system}} = 0 \tag{4.1}$$

Sign Convention Instead of specifying every time whether work is done on the system or by the system and whether heat is transferred to the system or from the system, let us adopt the following sign convention for W and Q (see Fig. 4.2). The work done by a system on the surroundings is treated as a positive quantity. Similarly, energy transfer as heat to the system from the surroundings is assigned a positive sign. With the sign convention Eq. (4.1) can be written as

$$\oint dQ = \oint dW \quad \text{or} \quad \oint (dQ - dW) = 0 \tag{4.2}$$

which is the First Law of Thermodynamics.

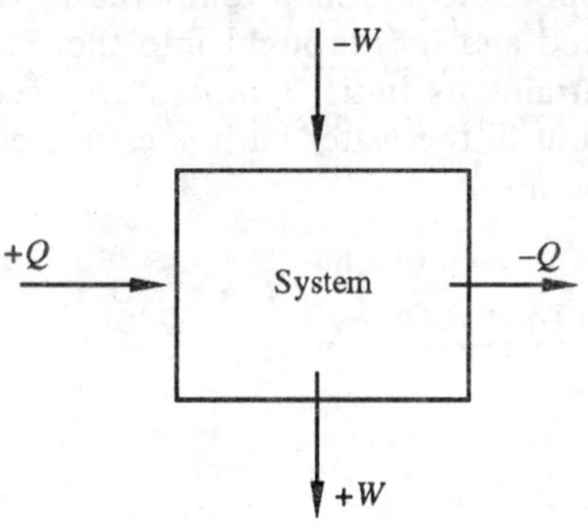

Fig. 4.2 Sign convention for heat and work interactions.

4.2 Consequences of the First Law of Thermodynamics

Heat is a Path Function Suppose a system is taken from state 1 to state 2 by the path 1-a-2 and is restored to the initial state by the path 2-b-1, then the system has undergone a cyclic process 1-a-2-b-1. If the system is restored to the initial state by the path 2-c-1, then the system has undergone the cyclic change 1-a-2-c-1 as shown in Fig. 4.3.

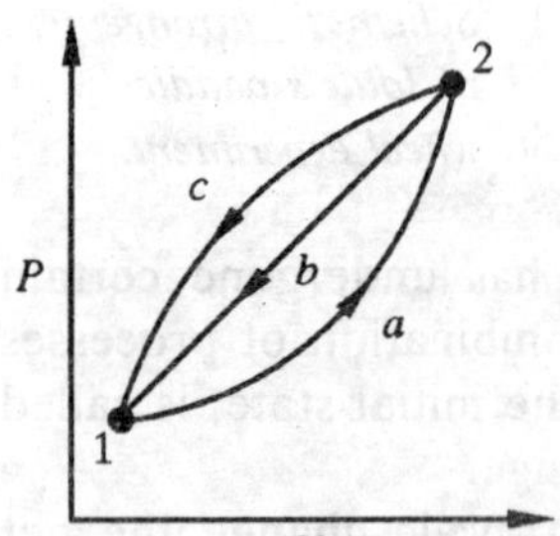

Fig. 4.3 Two cyclic processes on P–V diagram.

Let us apply the First Law of Thermodynamics to the cyclic processes 1-a-2-b-1 and 1-a-2-c-1 to obtain

$$\int_{1a2} dQ + \int_{2b1} dQ - \int_{1a2} dW - \int_{2b1} dW = 0 \tag{4.3}$$

$$\int_{1a2} dQ + \int_{2c1} dQ - \int_{1a2} dW - \int_{2c1} dW = 0 \tag{4.4}$$

Subtracting Eq. (4.4) from Eq. (4.3), we get

$$\int_{2b1} dQ - \int_{2c1} dQ - \left(\int_{2b1} dW - \int_{2c1} dW\right) = 0 \tag{4.5}$$

We know that work is a path function and hence

$$\int_{2b1} dW \neq \int_{2c1} dW$$

or $$\int_{2b1} dW - \int_{2c1} dW \neq 0 \tag{4.6}$$

From Eq. (4.5) and (4.6), we find

$$\int_{2b1} dQ \neq \int_{2c1} dQ \tag{4.7}$$

That is, heat is a path function.

Energy is a Property of the System Equation (4.5) can be rearranged as

$$\int_{2b1} (dQ - dW) = \int_{2c1} (dQ - dW)$$

which shows that $\int (dQ - dW)$ is the same for the paths 2-b-1 and 2-c-1, connecting the states 2 and 1. That is, the quantity $(dQ - dW)$ does not depend on the path followed by a system, but depends only on the initial and final states of the system. That is, $(dQ - dW)$ is an exact differential of a property. This property is called energy (E). It is given by

$$dE = dQ - dW \tag{4.8}$$

We know that a system possesses energy in several forms and the energy of the system is given by

$$E = \text{KE} + \text{PE} + U \tag{4.9}$$

where U is the internal energy. Therefore,

$$dE = d(\text{KE}) + d(\text{PE}) + dU = dQ - dW \tag{4.10}$$

Quite often an engineer deals with situations in which the KE and PE changes are negligible. In such cases the First Law of Thermodynamics (Eq. (4.10)) reduces to

$$dU = dQ - dW \tag{4.11}$$

Energy of an Isolated System is Conserved An isolated system does not exchange energy with its surroundings in the form of work as well as heat. Hence $dQ = 0$ and $dW = 0$. Then, the First Law of Thermodynamics (Eq. (4.8)) reduces to

$$dE = 0 \quad \text{or} \quad E_2 = E_1$$

That is, the energy of an isolated system remains constant. This fact is quite often referred to as the principle of conservation of energy.

Perpetual Motion Machine of the First Kind is Impossible An imaginary device which delivers work continuously without absorbing energy from its surroundings is called a Perpetual Motion Machine of the First Kind (PMMFK). Since the device has to deliver work continuously, it has to operate on a cycle. If such a device does not absorb energy from its surroundings $\oint dQ = 0$. From the First Law of Thermodynamics (Eq. (4.2)) it can be observed that $\oint dW = 0$, if $\oint dQ = 0$. That is, the First Law of Thermodynamics tells us that it is impossible to devise a PMMFK.

4.3 First Law Analysis of Non-flow Processes

The First Law of Thermodynamics can be applied to a system to evaluate the changes in its energy when it undergoes a change of state while interacting with its surroundings. The processes that are usually encountered in thermodynamic analysis of systems can be identified as any one or a combination of the following elementary processes:

- Constant volume (isochoric) process.
- Constant pressure (isobaric) process.
- Constant temperature (isothermal) process.
- Adiabatic process.

Application of the First Law of Thermodynamics to these processes is discussed below.

4.3.1 Constant Volume Process

Suppose a gas enclosed in a rigid vessel is interacting with the surroundings and absorbs energy Q as heat as shown in Fig. 4.4. Since the vessel is rigid, the work done W due to expansion or compression is zero. Applying the First Law of Thermodynamics, we get

$$dU = dQ \quad \text{or} \quad Q = U_2 - U_1 \tag{4.12}$$

That is, the heat interaction is equal to the change in the internal energy of the gas. If the system contains a mass m of an ideal gas, then

$$Q = \Delta U = mC_v(T_2 - T_1)$$

The path followed by the system is shown on a P–V diagram in Fig. 4.4.

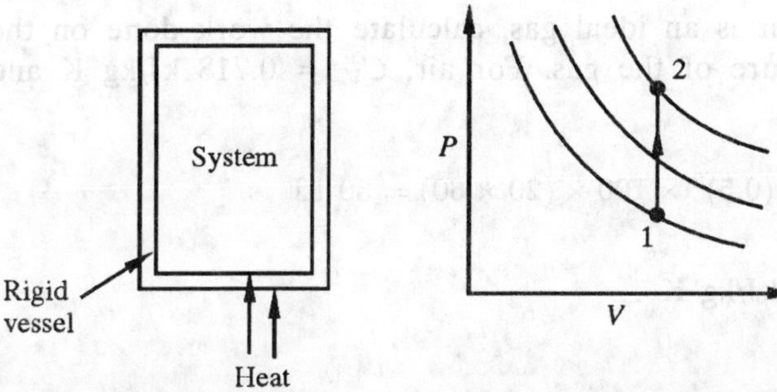

Fig. 4.4 Reversible constant volume process.

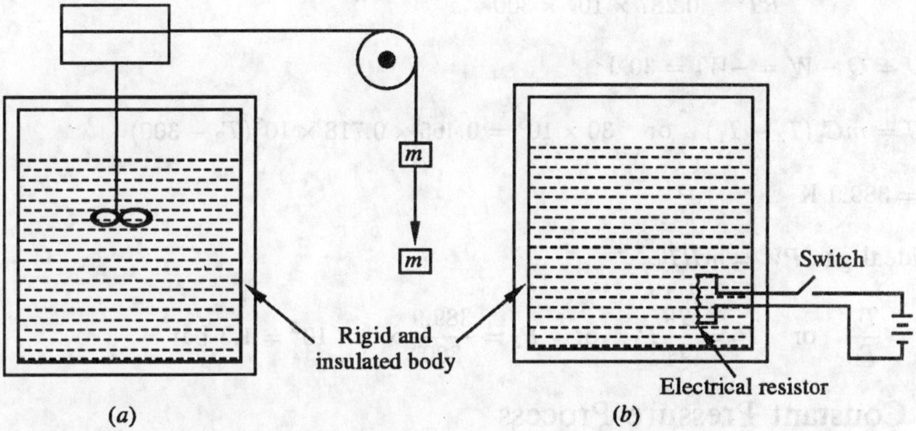

Fig. 4.5 (a) Shaft work done on a system at constant volume. (b) Constant volume process involving electrical work.

Now consider the fluid contained in a rigid vessel as shown in Fig. 4.5 as a system. The vessel is rigid and insulated. In Fig. 4.5(a), shaft work is done on the system by a paddle wheel and in Fig. 4.5(b), electrical work is done on the system. Since the vessel is rigid, the PdV work is zero. Moreover, the vessel is insulated and hence $dQ = 0$. Application of the First Law of Thermodynamics gives

$$dU = dQ - dW = dQ - (dW_{pdv} + dW_s)$$

or

$$dU = -dW_s \quad \text{or} \quad -W_s = \Delta U = U_2 - U_1 \tag{4.13}$$

where dW_{PdV} is the compression/expansion work and dW_s is the shaft work. That is, the increase in the internal energy of a system at constant volume, which is enclosed by an adiabatic wall, is equal to the shaft work done on the system.

EXAMPLE 4.1 An insulated and rigid container of 0.4 m^3 volume contains air at 1 bar and 300 K. A 100 Ω electrical resistor placed inside the container is supplied with 0.5 ampere current for 20 minutes. Assuming that air is an ideal gas, calculate the work done on the gas and the final pressure and temperature of the gas. For air, C_V = 0.718 kJ/kg K and M = 28.97 kg/kmol.

SOLUTION Work done, $-W_s = i^2Rt = (0.5)^2 \times 100 \times (20 \times 60) = 30$ kJ

For air, $R = \dfrac{R_u}{M} = \dfrac{8.314 \times 10^3}{28.97} = 0.287$ kJ/kg K

Mass of air, $m = \dfrac{PV}{RT} = \dfrac{1 \times 10^5 \times 0.4}{0.287 \times 10^3 \times 300} = 0.465$ kg

$$\Delta U = Q - W = -\dot{W}_s = 30 \text{ J}$$

$$\Delta U = mC_v(T_2 - T_1) \quad \text{or} \quad 30 \times 10^3 = 0.465 \times 0.718 \times 10^3(T_2 - 300)$$

or $T_2 = 389.9$ K

For an ideal gas $PV = mRT$

or $\dfrac{P_2}{P_1} = \dfrac{T_2}{T_1}$ or $P_2 = \dfrac{T_2}{T_1}P_1$ or $P_2 = \dfrac{389.9}{300} \times 1 \times 10^5 = 1.3$ bar

4.3.2 Constant Pressure Process

Several industrial processes are carried out at constant pressure. A few examples of constant pressure processes are shown in Fig. 4.6. The processes shown in Figs. 4.6(a) and 4.6(b) are reversible and the processes shown in Figs. 4.6(c) and 4.6(d) are irreversible.

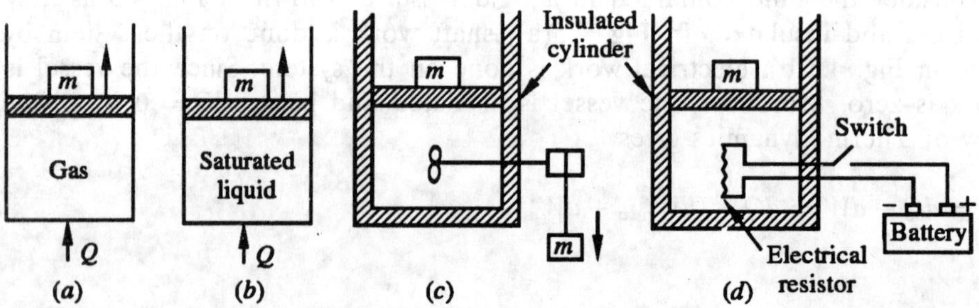

Fig. 4.6 Schematic arrangement of constant pressure processes involving (a) reversible heating/cooling of a gas. (b) phase change. (c) paddle wheel work. (d) electrical work.

For a constant pressure process, the work done W is given by

$$W = \int_1^2 PdV = P(V_2 - V_1) \tag{4.14}$$

Application of the First Law of Thermodynamics gives

$$dU = dQ - dW = dQ - PdV = dQ - d(PV)$$

or $$dQ = dU + d(PV) = d(U + PV) = dH$$

or $$Q = \Delta H \tag{4.15}$$

That is, in a constant pressure process, the heat interaction is equal to the increase in the enthalpy of the system.

Now, consider the constant pressure processes shown in Figs. 4.6(c) and 4.6(d) in which the system is enclosed by an adiabatic boundary. Application of the First Law of Thermodynamics gives

$$dU = dQ - dW = dQ - (PdV + dW_s) \tag{4.16}$$

Here, the work done (dW) consists of two parts—the PdV work associated with the motion of the boundary and ($-dW_s$), the shaft work (or electrical work) done by the surroundings. Since the system is enclosed by an adiabatic boundary, $dQ = 0$ and Eq. (4.16) can be rewritten as

$$-dW_s = dU + d(PV) = dH \tag{4.17}$$

That is, the increase in the enthalpy of the system is equal to the shaft work done on the system.

EXAMPLE 4.2 A piston-cylinder assembly contains 1 kg of saturated liquid water at 0.1 MPa. Energy is transferred as heat till all the liquid is converted into saturated vapor at the same pressure. Calculate the work done by water and the energy transferred as heat.

Specific volume of saturated liquid at 0.1 MPa $= v_f = 1.0432 \times 10^{-3}$ m³/kg

Specific volume of saturated vapor at 0.1 MPa $= v_g = 1694 \times 10^{-3}$ m³/kg

Specific enthalpy of saturated vapor at 0.1 MPa $= h_f = 417.46$ kJ/kg

Specific enthalpy of saturated vapor at 0.1 MPa $= h_g = 2675.5$ kJ/kg

SOLUTION Work done, $W = P(V_2 - V_1) = 0.1 \times 10^6 (1694 - 1.0432) \times 10^{-3} = 169.3$ kJ

For a constant pressure process,

$$q = \Delta h = h_g - h_f = 2675.5 - 417.46 = 2258.04 \text{ kJ}$$

EXAMPLE 4.3 The densities of ice and water at 0°C are 916.23 kg/m³ and 999.8 kg/m³, respectively. The latent heat of fusion of water at 0°C is 334.98 kJ. If 1 kg of ice melts at 0°C, calculate the work done and the change in internal energy.

SOLUTION Ice melts into water at 0°C and 1 bar pressure. Specific volume of ice,

$$v_s = \frac{1}{916.23} = 1.0914 \times 10^{-3} \text{ m}^3/\text{kg}$$

Specific volume of water, $v_f = \dfrac{1}{999.8} = 1.0002 \times 10^{-3} \text{ m}^3/\text{kg}$

$$W = P(v_f - v_s) = 1 \times 10^5 (1.0002 - 1.0914) \times 10^{-3} = -9.12 \text{ J}$$

For a constant pressure process, $q = \Delta h = 334.98$

$$\Delta h = \Delta u + \Delta(Pv) = \Delta u + P\Delta v = \Delta u + W$$

$$\Delta u = \Delta h - W = 334.98 \times 10^3 - (-9.12) = 334.99 \text{ kJ}$$

4.3.3 Constant Temperature Process

Suppose a gas enclosed in a piston-cylinder assembly is allowed to expand from P_1 to P_2 while the temperature is held constant. Then application of the First Law of Thermodynamics gives

$$dU = dQ - dW = dQ - PdV$$

It is not possible to calculate the heat and work interactions unless the relationships between the thermodynamic properties of the gas are known. Suppose the gas under consideration is an ideal gas (which follows the relation $Pv = RT$ and $u = u(T)$ only) then for an isothermal process.

$$dU = 0$$

$$dQ = Pdv = RT\frac{dv}{v} \quad \text{or} \quad Q = W = RT \ln\left(\frac{v_2}{v_1}\right)$$

EXAMPLE 4.4 Calculate the work to be done on 1 kmol of ammonia when it is isothermally compressed from the initial state 5 m³/kmol to 0.5 m³/kmol at 300 K. It is known that ammonia obeys the van der Waals equation of state. The van der Waals constants for ammonia are $a = 423.3$ kPa(m³/kmol)² and $b = 0.0373$ m³/kmol.

SOLUTION $W = \displaystyle\int_1^2 Pdv = \int_1^2 \left(\frac{RT}{v-b} - \frac{a}{v^2}\right) dv = RT \ln\frac{v_2 - b}{v_1 - b} - a\left(\frac{1}{v_1} - \frac{1}{v_2}\right)$

$$= 8.314 \times 300 \ln\left(\frac{0.5 - 0.0373}{5 - 0.0373}\right) - 423.3\left(\frac{1}{5} - \frac{1}{0.5}\right) = -5155.86 \text{ kJ}$$

4.3.4 Reversible Adiabatic (Isentropic) Process

Suppose a gas contained in a piston-cylinder assembly which is insulated is allowed to expand reversibly from the initial pressure P_1 to the final pressure P_2. Application of the First Law of Thermodynamics gives

$$dU = -dW \quad \text{or} \quad W = -\Delta U \tag{4.18}$$

That is, in an adiabatic process, the system does work at the expense of its own internal energy. Equation (4.18) is true for reversible as well as irreversible, adiabatic processes. Now, let us consider an ideal gas which is undergoing a reversible, adiabatic process. For an ideal gas, $Pv = RT$ and $u = u(T)$ only. Equation (4.18) can be rewritten as

$$C_v dT = -P dv = -\frac{RT}{v} dv$$

$$\text{or} \quad \frac{dT}{T} = -\frac{R}{C_v}\frac{dv}{v} \tag{4.19}$$

For an ideal gas, we know that

$$C_p - C_v = R \quad \text{or} \quad \frac{C_p - C_v}{C_v} = \gamma - 1 = \frac{R}{C_v}$$

Therefore, Eq. (4.19) reduces to

$$\frac{dT}{T} = -(\gamma - 1)\frac{dv}{v} \quad \text{or} \quad \ln\frac{T_2}{T_1} = -(\gamma - 1)\ln\frac{v_2}{v_1} = \ln\left(\frac{v_1}{v_2}\right)^{\gamma-1}$$

or

$$\frac{T_2}{T_1} = \left(\frac{v_1}{v_2}\right)^{\gamma-1} \quad \text{or} \quad Tv^{\gamma-1} = \text{constant} \tag{4.20}$$

Equation (4.20) can also be written as

$$\frac{T_2}{T_1} = \left(\frac{RT_1}{P_1}\frac{P_2}{RT_2}\right)^{\gamma-1} = \left(\frac{P_2}{P_1}\right)^{\gamma-1}\left(\frac{T_1}{T_2}\right)^{\gamma-1}$$

$$\frac{T_2}{T_1}\left(\frac{T_2}{T_1}\right)^{\gamma-1} = \left(\frac{T_2}{T_1}\right)^{\gamma} = \left(\frac{P_2}{P_1}\right)^{\gamma-1}$$

$$\text{or} \quad \frac{T_2}{T_1} = \left(\frac{P_2}{P_1}\right)^{(\gamma-1)/\gamma} \tag{4.21}$$

Equation (4.20) can also be written as

$$\frac{T_2}{T_1} = \frac{P_2 v_2}{R} \times \frac{R}{P_1 v_1} = \left(\frac{P_2}{P_1}\right)\left(\frac{v_2}{v_1}\right) = \left(\frac{v_1}{v_2}\right)^{\gamma-1}$$

or

$$\frac{P_2}{P_1} = \left(\frac{v_1}{v_2}\right)^{\gamma} \quad \text{or} \quad Pv^{\gamma} = \text{constant} \tag{4.22}$$

EXAMPLE 4.5 In an automobile engine, air (an ideal gas with $\gamma = 1.4$) at 1 bar and 300 K is compressed to one-eighth of its original volume, in a reversible adiabatic process. Calculate the final temperature and pressure of the air and the work done per kg air.

SOLUTION It is given that $v_1/v_2 = 8$. For an ideal gas undergoing a reversible adiabatic process,

$$P_1 v_1^{\gamma} = P_2 v_2^{\gamma} \quad \text{or} \quad P_2 = P_1 \left(\frac{v_1}{v_2}\right)^{\gamma} = 1 \times (8)^{1.4} = 18.38 \text{ bar}$$

$$\frac{T_2}{T_1} = \left(\frac{v_1}{v_2}\right)^{\gamma-1}$$

$$T_2 = T_1 \left(\frac{v_1}{v_2}\right)^{\gamma-1} = 300(8)^{1.4-1} = 689.2 \text{ K}$$

$$W = \frac{P_1 v_1 - P_2 v_2}{\gamma - 1} = \frac{R(T_1 - T_2)}{\gamma - 1} = \frac{8.314 \times (300 - 689.2)}{1.4 - 1} = -8089.5 \text{ J/mol}$$

$$\equiv -\frac{8089.5}{28.97 \times 10^{-3}}$$

$$= 2279.24 \text{ kJ/kg}$$

4.3.5 Polytropic Process

If a gas follows the path Pv^n = constant during expansion or compression, the process is called a polytropic process. If a gas changes its state from (P_1, v_1) to (P_2, v_2) following a polytropic process, the work done by the gas is given by

$$W = \frac{P_1 v_1 - P_2 v_2}{n - 1} \tag{4.23}$$

Applying the First Law of Thermodynamics for a polytropic process, we get

$$dU = dQ - dW$$

or

$$U_2 - U_1 = Q - \frac{P_1 v_1 - P_2 v_2}{n - 1} \tag{4.24}$$

Depending on the value of n (the index of expansion), a polytropic process reduces to the following common processes shown in Fig. 4.7:

$n = 0$; Constant pressure process

$n = 1$; Constant temperature process

$n = \gamma$; Reversible adiabatic process

$n = \infty$; Constant volume process

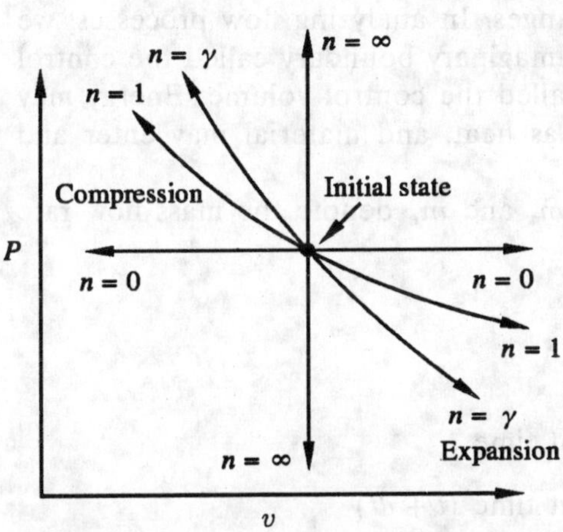

Fig. 4.7 Polytropic process on P–v diagram.

EXAMPLE 4.6 Air (ideal gas with $\gamma = 1.4$) at 1 bar and 300 K is compressed till the final volume is one-sixteenth of the original volume, following a polytropic process $Pv^{1.25}$ = constant. Calculate (a) the final pressure and temperature of the air, (b) the work done and (c) the energy transferred as heat per mole of air.

SOLUTION (a) $P_1 v_1^{1.25} = P_2 v_2^{1.25}$

$$P_2 = P_1 \left(\frac{v_1}{v_2}\right)^{1.25} = 1(16)^{1.25} = 32 \text{ bar}$$

$$T_2 = \frac{T_1 P_2 v_2}{P_1 v_1} = 300 \times \frac{32}{1} \times \frac{1}{16} = 600 \text{ K}$$

(b) $$W = \frac{P_1 v_1 - P_2 v_2}{n - 1} = \frac{R(T_1 - T_2)}{n - 1} = \frac{8.314(300 - 600)}{1.25 - 1} = -9.977 \text{ kJ/mol}$$

(c) $\Delta u = q - W$

$$c_v(T_2 - T_1) = \frac{R}{\gamma - 1}(T_2 - T_1) = q - W$$

$$q = \frac{R(T_2 - T_1)}{\gamma - 1} + W = \frac{8.314 \times (600 - 300)}{1.4 - 1} - 9.977 \times 10^3$$

or $\quad q = -3.742$ kJ/mol

4.4 First Law Analysis of Flow Processes

In the first law analysis of non-flow processes, attention is focused on a definite quantity of matter (or system). Quite often engineers deal with devices like turbines, compressors, pumps, etc., through which material flows continuously. It is quite difficult to follow the course of each element of matter and to evaluate the energy changes. In analyzing flow processes, we focus our attention on the device and enclose it by an imaginary boundary called the control surface. The space bounded by the control surface is called the control volume. Energy may cross the control surface in the form of work as well as heat, and material may enter and leave the control volume (CV).

Consider the control volume shown in Fig. 4.8. Let $\dot{m}_i$ and $\dot{m}_e$ denote the mass flow rate at the inlet i and exit e of the control volume. Let

$m(t)$ = mass inside the CV at time t

$m(t + dt)$ = mass inside the CV at time $(t + dt)$

$e(t)$ = specific energy of mass in the CV at time t

$e(t + dt)$ = specific energy of mass in the CV at time $(t + dt)$

$e_i = (u_i + \mathbf{V}_i^2/2 + gz_i)$ = specific energy of mass at inlet i

e_e = specific energy of mass at exit e

$\mathbf{V}_i$ = flow velocity of mass at inlet i

$\mathbf{V}_e$ = flow velocity of mass at exit e

P_i = pressure at inlet i

P_e = pressure at exit e

v_i = specific volume at inlet i

v_e = specific volume at exit e

z_i = elevation at inlet i

z_e = elevation at exit e

$\dot{Q}$ = rate of heat flow into the CV

$\dot{W}_s$ = rate of shaft work done by CV

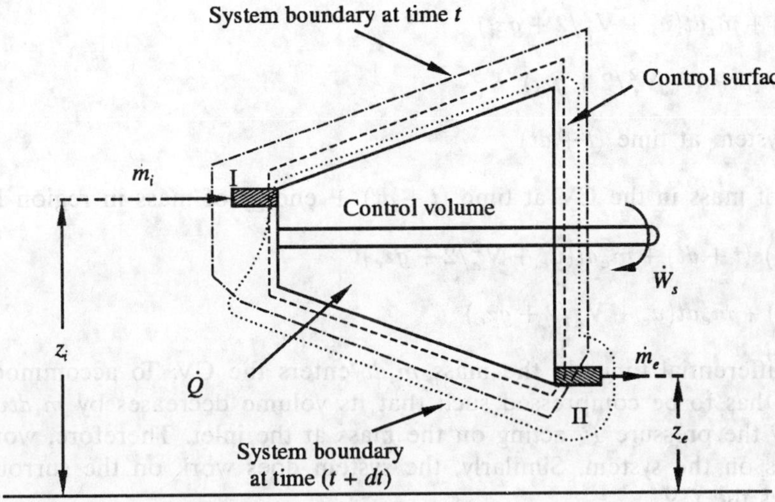

Fig. 4.8 Schematic representation of control volume.

Consider a differential time dt while the matter flows through the control volume. During this time,

Mass entering the control volume $= \dot{m}_i dt$

Mass leaving the control volume $= \dot{m}_e dt$

Then, the conservation of mass gives

$$m(t+dt) = m(t) + \dot{m}_i dt - \dot{m}_e dt$$

or
$$\frac{m(t+dt) - m(t)}{dt} = \frac{dm}{dt} = \dot{m}_i - \dot{m}_e \tag{4.25}$$

In order to apply the First Law of Thermodynamics, let us choose the system such that the mass of the system remains constant all the time. At time t we choose the system as the mass in the CV and the mass in region I which enters the CV in time dt. During the time dt, the mass in region I enters the CV and some mass leaves the CV and occupies region II as shown in Fig. 4.8. During this time,

Heat flow into the system $= \dot{Q}dt$

Shaft work done by the system $= \dot{W}_s dt$

Energy of the system at time t

= energy of mass in the CV at time t + energy of mass in region I

$$= m(t)e(t) + \dot{m}_i dt(u_i + \mathbf{V}_i^2/2 + gz_i)$$

$$= E(t) + \dot{m}_i dt(u_i + \mathbf{V}_i^2/2 + gz_i)$$

Energy of the system at time $(t + dt)$

$$= \text{energy of mass in the CV at time } (t + dt) + \text{energy of mass in region II}$$

$$= m(t + dt)e(t + dt) + \dot{m}_e dt(u_e + \mathbf{V}_e^2/2 + gz_e)$$

$$= E(t + dt) + \dot{m}_e dt(u_e + \mathbf{V}_e^2/2 + gz_e)$$

During the differential time dt, the mass $\dot{m}_i dt$ enters the CV. To accommodate this, the mass in the CV has to be compressed such that its volume decreases by $\dot{m}_i dt v_i$, and this is accomplished by the pressure P_i acting on the mass at the inlet. Therefore, work is done by the surroundings on the system. Similarly, the system does work on the surroundings when mass $\dot{m}_e dt$ leaves the CV.

$$\text{Total work done due to flow} = P_e \dot{m}_e dt v_e - P_i \dot{m}_i dt v_i$$

We know that the First Law of Thermodynamics is given by

$$\Delta E = Q - W$$

Substituting the various terms in the above expression, we get

$$[E(t + dt) + \dot{m}_e dt(u_e + \mathbf{V}_e^2/2 + gz_e)] - [E(t) + \dot{m}_i dt(u_i + \mathbf{V}_i^2/2 + gz_i)]$$

$$= \dot{Q}dt - [\dot{W}_s dt + P_e \dot{m}_e dt v_e - P_i \dot{m}_i dt v_i]$$

$$\frac{E(t + dt) - E(t)}{dt} = \dot{m}_i(u_i + P_i v_i + \mathbf{V}_i^2/2 + gz_e)] - \dot{m}_e(u_e + P_e v_e + \mathbf{V}_e^2/2 + gz_e) + \dot{Q} - \dot{W}_s$$

$$\frac{dE}{dt} = \dot{m}_i(h_i + V_i^2/2 + gz_i) - \dot{m}_e(h_e + V_e^2/2 + gz_e) + \dot{Q} - \dot{W}_s \tag{4.26}$$

which is the most general form of the First Law of Thermodynamics for a control volume.

4.5 Steady-State, Steady-Flow Processes

Many of the devices used in industries operate continuously and in steady-state. Therefore, an analysis of steady-state flow processes is useful in designing and evaluating the performance of such devices. If the equipment operates under steady-state conditions, there is no accumulation of mass and energy in the control volume and the rate of shaft work done and heat flow are constant with time. In other words, steady-state steady-flow implies that

$$\dot{m}_i = \dot{m}_e = \dot{m} \quad \text{and} \quad \frac{dE}{dt} = 0$$

$$\dot{Q} = \text{constant}; \qquad \dot{W}_s = \text{constant}$$

For steady-state, steady-flow conditions, the First Law of Thermodynamics (Eq. (4.26)) reduces to

$$\left(h_i + \mathbf{V}_i^2/2 + gz_i\right) = \left(h_e + \mathbf{V}_e^2/2 + gz_e\right) = \frac{\dot{W}_s - \dot{Q}}{\dot{m}} \tag{4.27}$$

Quite often, the shaft work done by a control volume is called the mechanical work done in a steady-flow process.

EXAMPLE 4.7 A turbine in a steam power plant operating under steady-state conditions receives superheated steam at 3 MPa and 320°C ($h = 3043.4$ kJ/kg) at a rate of 1 kg/s. The steam enters the turbine with a velocity of 10 m/s at an elevation of 5 m above the ground level. The wet steam at 0.1 bar (h = 2250 kJ/kg) leaves the turbine at a velocity of 40 m/s at an elevation of 10 m above the ground level. Calculate the power output of the turbine.

SOLUTION Given data:

$$h_i = 3043.4 \text{ kJ/kg}; \ \ \mathbf{V}_i = 10 \text{ m/s}; \ \ z_i = 5 \text{ m}$$

$$h_e = 2050 \text{ kJ/kg}; \ \ \mathbf{V}_e = 40 \text{ m/s}; \ \ z_e = 10 \text{ m}; \ \ \dot{Q} = 0; \ \ \dot{m} = 1 \text{ kg/s}$$

The First Law of Thermodynamics for steady-state, steady-flow process is given by

$$\left(h_i + \mathbf{V}_i^2/2 + gz_i\right) - \left(h_e + \mathbf{V}_e^2/2 + gz_e\right) = \frac{\dot{W}_s - \dot{Q}}{\dot{m}}$$

$$(h_i - h_e) + \frac{\mathbf{V}_i^2 - \mathbf{V}_e^2}{2} + g(z_i - z_e) = \frac{\dot{W}_s - \dot{Q}}{\dot{m}}$$

Substituting the given data, we get

$$(3043.4 - 2250) \times 10^3 + \left(\frac{10^2 - 40^2}{2}\right) + 9.81(5 - 10) = \dot{W}_s$$

or $\quad 793.4 \times 10^3 - 750 - 49 = \dot{W}_s \quad$ or $\quad \dot{W}_s = 792.6$ kW

EXAMPLE 4.8 If the changes in KE and PE are ignored in Example 4.7, estimate the percent error.

SOLUTION Change in KE = 750 W (see Example 4.7). Change in PE = 49 W. Change in enthalpy = 793.4 kW.

$$\text{Percent error} = \frac{750 + 49}{793.4 \times 10^3} \times 100 = 0.1$$

4.6 Applications of Steady-Flow Processes

Turbine High pressure steam enters a turbine and after expansion leaves the turbine at low pressure. It was shown in Example 4.8 that the changes in the KE and PE of the steam while passing through the turbine is negligible. Hence, it is quite reasonable to ignore the changes in KE and PE while analyzing the performance of a turbine. If the turbine is adiabatic the power output of the turbine is given by

$$\dot{W}_s = \dot{m}(h_i - h_e) \tag{4.28}$$

Compressor A compressor receives a gas at low pressure and delivers the gas at a high pressure when work is done on the unit. It is reasonable to ignore the changes in KE and PE of the gas flowing through the compressor. The First Law of Thermodynamics for an adiabatic compressor operating under steady-state conditions, gives the power input to the compressor as

$$-\dot{W}_s = \dot{m}(h_e - h_i) \tag{4.29}$$

Nozzle A nozzle, shown in Fig. 4.9, is a device to increase the flow velocity. Since the objective of the device it to convert the energy into KE, one cannot ignore the change in KE, but one can ignore the change in PE. For an adiabatic nozzle operating at steady-state conditions, the First Law of Thermodynamics reduces to

$$h_i + \mathbf{V}_i^2/2 = h_e + \mathbf{V}_e^2/2$$

If the inlet velocity of the fluid is negligible ($\mathbf{V}_i = 0$), the exit velocity of the fluid is given by

$$\mathbf{V}_e = \sqrt{2(h_i - h_e)} \tag{4.30}$$

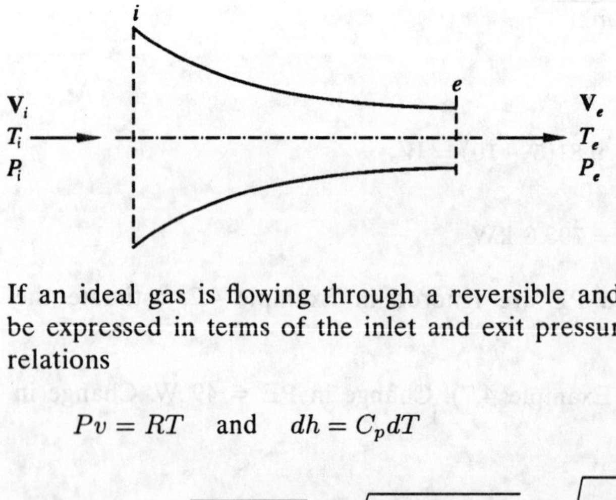

Fig. 4.9 *Schematic representation of a nozzle.*

If an ideal gas is flowing through a reversible and adiabatic nozzle, the exit velocity $\mathbf{V}_e$ can be expressed in terms of the inlet and exit pressures and temperatures by making use of the relations

$$Pv = RT \quad \text{and} \quad dh = C_p dT$$

Then $$\mathbf{V}_e = \sqrt{2(h_i - h_e)} = \sqrt{2C_p(T_i - T_e)} = \sqrt{2C_pT_i\left(1 - \frac{T_e}{T_i}\right)} \tag{4.31}$$

For an ideal gas undergoing adiabatic expansion

$$\frac{T_e}{T_i} = \left(\frac{P_e}{P_i}\right)^{(\gamma-1)/\gamma}$$

Therefore,

$$\mathbf{V}_e = \sqrt{2C_pT_i\left[1-\left(\frac{P_e}{P_i}\right)^{(\gamma-1)/\gamma}\right]} \tag{4.32}$$

EXAMPLE 4.9 Superheated steam at 3 bar and 160°C (h = 2782.4 kJ/kg) enters an adiabatic nozzle with negligible velocity and leaves at 1 bar and 120°C (h = 2716.6 kJ/kg). Calculate the discharge velocity of steam.

SOLUTION The exit velocity is given by

$$\mathbf{V}_e = \sqrt{2(h_i - h_e)} = \sqrt{2(2782.3 - 2716.6)\times 10^3} = 362.5 \text{ m/s}$$

EXAMPLE 4.10 Air (ideal gas, γ = 1.4) expands reversibly and adiabatically through a nozzle from inlet conditions of 3 bar and 400 K to an exit pressure of 1 bar. Calculate the exit velocity of the air.

SOLUTION Given data: P_i = 3 bar; T_i = 400 K; and P_e = 1 bar

For air, $R = \frac{R_u}{M} = \frac{8.314\times 10^3}{28.97} = 0.287$ kJ/kg K

$$C_p = \frac{R\gamma}{\gamma-1} = \frac{0.287\times 1.4}{1.4-1} = 1.0045 \text{ kJ/kg K}$$

The exit velocity of air is given by

$$\mathbf{V}_e = \sqrt{2C_pT_i\left[1-\left(\frac{P_e}{P_i}\right)^{(\gamma-1)/\gamma}\right]}$$

Substituting the values in the above equation, we get

$$\mathbf{V}_e = \sqrt{2\times 1.0045\times 10^3\times 400\left[1-\left(\frac{1}{3}\right)^{0.4/1.4}\right]} = 465.3 \text{ m/s}$$

Condensers/Heat Exchangers The primary objective of a heat exchanger or a condenser is to transfer heat from a hot fluid to a cold fluid. A schematic of a condenser/heat exchanger is shown in Fig. 4.10.

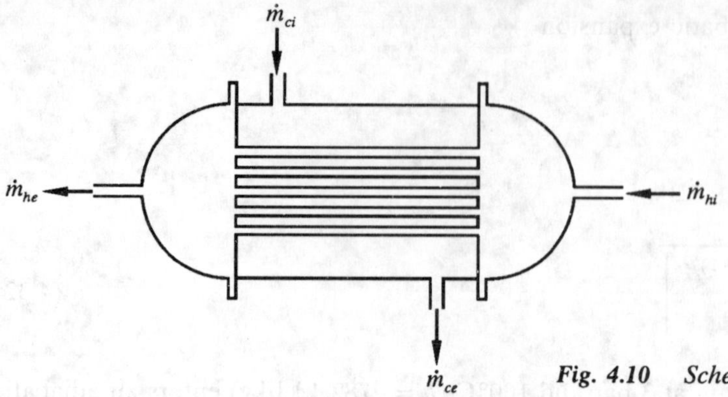

Fig. 4.10 *Schematic of a heat exchanger/condenser.*

Consider the condenser/heat exchanger as the control volume. This control volume has 2 inlets and 2 exits. The heat exchanger/condenser operates under steady-state conditions. For steady-state, steady-flow conditions, the First Law of Thermodynamics (Eq. (4.26)) can be written as

$$\sum \dot{m}_i(h_i + \mathbf{V}_i^2/2 + gz_i) - \sum \dot{m}_e(h_e + \mathbf{V}_e^2/2 + gz_e) + \dot{Q} - \dot{W}_s = 0 \tag{4.33}$$

where the summation is to be carried over all inlet and exit ports. In a condenser/ heat exchanger, it is reasonable to ignore changes in KE and PE terms. No shaft work is delivered by the control volume and hence $\dot{W}_s = 0$. If there are no heat losses to the surroundings, $\dot{Q} = 0$. Then Eq. (4.33) reduces to

$$\dot{m}_{hi}h_{hi} + \dot{m}_{ci}h_{ci} - \dot{m}_{he}h_{he} - \dot{m}_{ce}h_{ce} = 0 \tag{4.34}$$

The conservation of mass gives

$$\dot{m}_{hi} = \dot{m}_{he} = \dot{m}_h$$

$$\dot{m}_{ci} = \dot{m}_{ce} = \dot{m}_c$$

Therefore, Eq. (4.34) can be rearranged as

$$\dot{m}_h(h_{hi} - h_{he}) = \dot{m}_c(h_{ce} - h_{ci}) \tag{4.35}$$

where the subscripts h and c denote the hot and cold fluids, respectively. ***Boilers*** The primary objective of a boiler is to produce steam from water by supplying energy as heat. A schematic of a boiler is shown in Fig. 4.11.

In a boiler, no shaft work is involved ($\dot{W}_s = 0$). It is reasonable to ignore the changes in KE and PE terms. If the boiler is operating under steady-state conditions, the First Law of Thermodynamics reduces to

$$\dot{m}(h_i - h_e) + \dot{Q} = 0 \quad \text{or} \quad \dot{Q} = \dot{m}(h_e - h_i) \tag{4.36}$$

That is, the rate at which heat is to be supplied in a boiler is equal to the product of mass flow rate and the increase in the enthalpy of the water.

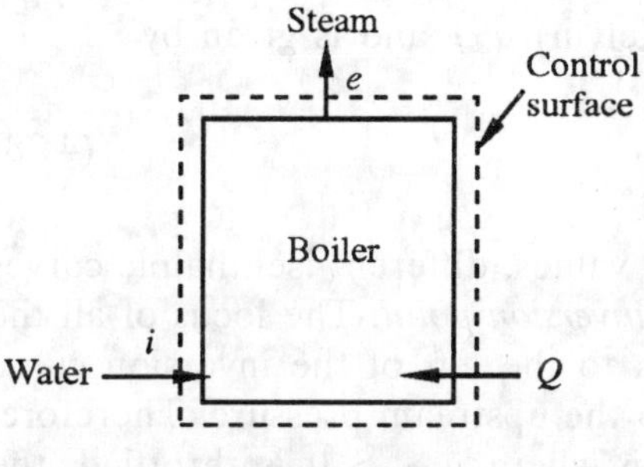

Fig. 4.11 Schematic of a boiler.

4.7 Throttling Process

Joule and Thomson (also known as Lord Kelvin) conducted the porous plug experiment in 1853 to measure the temperature change when a fluid is throttled from high pressure to low pressure. Consider the steady-state, steady-flow of a fluid through a long, horizontal and insulated pipe in which some portion of the pipe is packed with a porous material as shown in Fig. 4.12. The fluid at pressure P_i and temperature T_i enters the control volume and leaves at pressure P_e and temperature T_e. The pipe is horizontal ($z_i = z_e$) and insulated ($\dot{Q} = 0$). The control volume does not deliver any shaft work ($\dot{W}_s = 0$). Under these conditions the First Law of Thermodynamics, (Eq. (4.27)) reduces to

$$h_i = h_e \tag{4.37}$$

That is, the throttling process is a constant enthalpy (or isenthalpic) process. Therefore, whenever a fluid flows through a porous plug or a partially opened valve and expands from high pressure to low pressure, its enthalpy remains constant and the fluid is said to have undergone a throttling process.

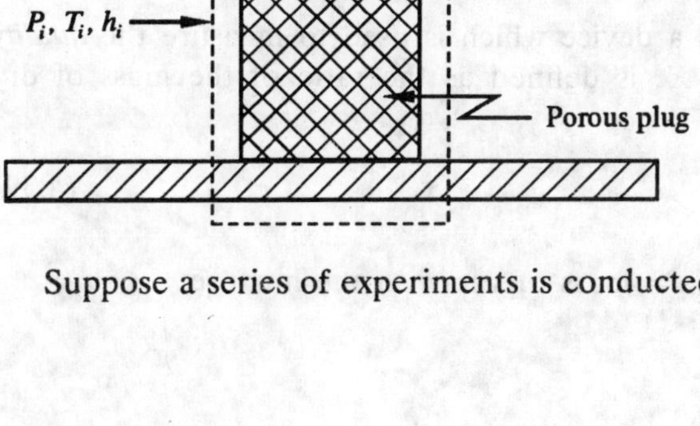

Fig. 4.12 Sketch of porous plug experiment.

Suppose a series of experiments is conducted on the same gas with the same initial pressure

P_i, one would observe different temperatures T_e. Since P_i and T_i are held constant, the enthalpy h_i is constant and the final set of P_e and T_e correspond to the same enthalpy. A plot of T_e versus P_e yields the isenthalpic curve as shown in Fig. 4.13. The slope of the isenthalpic curve is called the Joule–Thomson coefficient (or Joule–Kelvin) μ_{JT} and is given by

$$\mu_{JT} = \left(\frac{\partial T}{\partial P}\right)_h \tag{4.38}$$

If the experiments are repeated with different P_i and T_i values, different isenthalpic curves will be obtained. The point at which $\mu_{JT} = 0$ is called the *inversion point*. The locus of all the inversion points is called the *inversion curve*. In the region to the left of the inversion curve $\mu_{JT} > 0$. In throttling, the downstream pressure is less than the upstream pressure. Therefore, whenever a real gas, the state of which lies in the region where $\mu_{JT} > 0$ is throttled, the temperature of the gas decreases. In the region to the right of the inversion curve $\mu_{JT} < 0$. If a real gas, the initial state of which lies the region $\mu_{JT} < 0$ is throttled, the temperature of the gas increases. For an ideal gas $\mu_{JT} = 0$ and hence there will be no change in the temperature if an ideal gas is throttled.

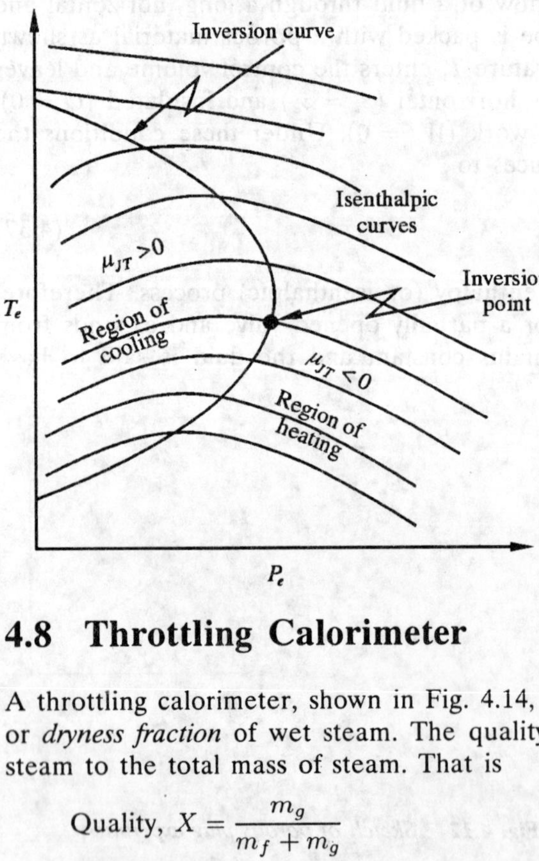

Fig. 4.13 *Inversion and isenthalpic curves.*

4.8 Throttling Calorimeter

A throttling calorimeter, shown in Fig. 4.14, is a device which is used to measure the *quality* or *dryness fraction* of wet steam. The quality, X, is defined as the ratio of the mass of dry steam to the total mass of steam. That is

$$\text{Quality, } X = \frac{m_g}{m_f + m_g} \tag{4.39}$$

where m_f = mass of liquid in the mixture; m_g = mass of vapor in the mixture.

The throttling calorimeter is based on the principle of Joule–Thomson expansion. A small amount of wet steam flowing through a steam main is throttled to a low pressure, such that the exit steam, is superheated. Since throttling is an isenthalpic process, the enthalpy of the wet steam is equal to the enthalpy of superheated steam leaving the throttling calorimeter. By measuring the pressure and temperature of the superheated steam, its enthalpy can be estimated or read from the steam tables. In the wet steam, the saturated liquid as well as saturated vapor coexist in a state of equilibrium and hence pressure and temperature are related to each other. Therefore, a knowledge of pressure and temperature is not enough to determine the quality of wet steam. The method of determining the quality of wet steam is discussed in chapter 6 while discussing the properties of steam.

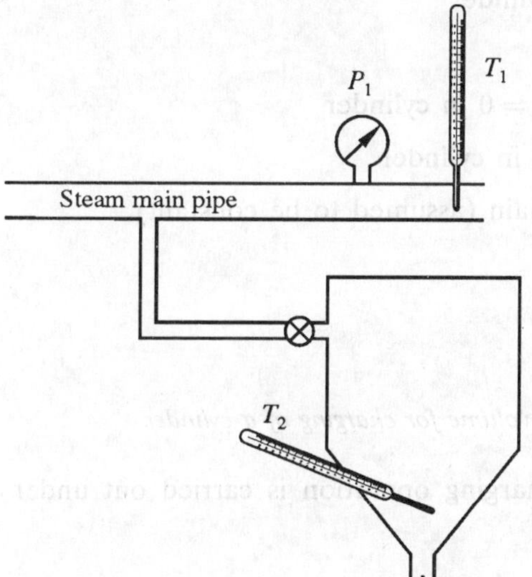

Fig. 4.14 Schematic of a throttling calorimeter.

4.9 Transient or Unsteady-State Flow Processes

Many processes of interest to an engineer are carried out in steady-state, steady-flow conditions. However, at times an engineer has to deal with unsteady-state or transient flow processes also. Examples of unsteady-state flow processes are charging of a cylinder and discharging of a cylinder. The two special cases of transient flow processes, namely charging of a cylinder and discharging of a cylinder are presented below.

Charging of a Cylinder Consider the cylinder shown in Fig. 4.15 as the control volume. The general expression for the First Law of Thermodynamics to a control volume is given by

$$\frac{dE}{dt} = \dot{m}_i(h_i + \mathbf{V}_i^2/2 + gz_i) - \dot{m}_e(h_e + \mathbf{V}_e^2/2 + gz_e) + \dot{Q} - \dot{W}_s \quad (4.26)$$

While charging a cylinder, no material leaves the control volume and no shaft work is done by the control volume. Hence $\dot{m}_e = 0$ and $\dot{W}_s = 0$. Ignore the KE and PE terms. Then Eq. (4.26) reduces to

$$\frac{dE}{dt} = \dot{m}_i h_i + \dot{Q} \quad \text{or} \quad \int_0^t \dot{m}_i h_i dt + \int_0^t \dot{Q} dt = \int_0^t \frac{dE}{dt} dt$$

or

$$h_i(m_f - m_0) + Q = E_f - E_0 = m_f u_f - m_0 u_0 \tag{4.40}$$

where m_0 = initial mass (at time $t = 0$) of gas in cylinder

m_f = final mass (at time t) of gas in cylinder

u_0 = specific internal energy of gas at time $t = 0$ in cylinder

u_f = specific internal energy of gas at time t in cylinder

h_i = specific enthalpy of gas in the supply main (assumed to be constant)

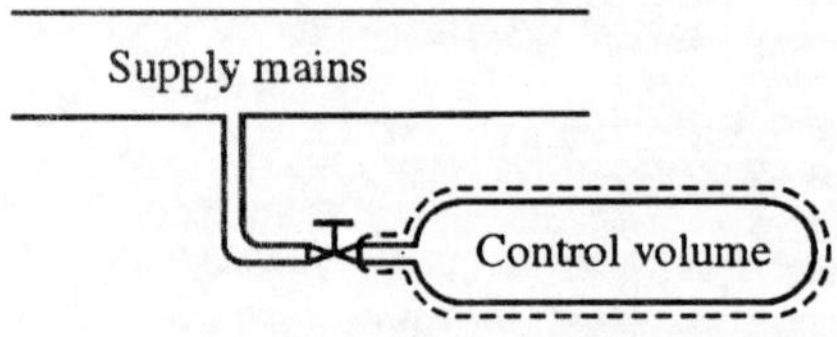

Fig. 4.15 Control volume for charging of a cylinder.

If the cylinder is initially empty ($m_0 = 0$) and the charging operation is carried out under adiabatic conditions ($Q = 0$), Eq. (4.40) reduces to

$$h_i = u_f \tag{4.41}$$

That is, the final internal energy of the gas in the cylinder is equal to the enthalpy of the gas in the supply main.

If an ideal gas is charged into an empty cylinder under adiabatic conditions, Eq. (4.41) further reduces to

$$C_p T_i = C_v T_f$$

or $$T_f = \gamma T_i \tag{4.42}$$

where T_i and T_f denote the temperature of the gas in the supply main and in the cylinder at the end of the charging operation.

EXAMPLE 4.11 An evacuated and insulated tank of 1 m^3 volume is connected to a line carrying air (ideal gas, $\gamma = 1.4$) at 20 bar and 300 K and the air is filled in the tank to a pressure of 20 bar. Calculate the final temperature of air in the tank and the mass of air in the tank at the end of the filling operation.

SOLUTION If an ideal gas is charged into an evacuated and insulated tank, the final temperature of the gas is given by

$$T_f = \gamma T_i = 1.4 \times 300 = 420 \text{ K}$$

$$R(\text{air}) = \frac{R_u}{M} = \frac{8.314 \times 10^3}{28.97} = 0.287 \text{ kJ/kg K}$$

$$\text{Mass of air, } m = \frac{PV}{RT} = \frac{20 \times 10^5 \times 1}{0.287 \times 10^3 \times 420} = 16.59 \text{ kg}$$

EXAMPLE 4.12 An insulated tank of 1 m³ capacity contains methane at 1 bar and 300 K. Attached to this tank is a line in which methane at 20 bar and 300 K is flowing. Methane from this line enters the tank till the methane pressure in the tank rises to 20 bar. Assuming that methane behaves like an ideal gas, determine the final temperature of methane and the mass of methane which entered the tank. For methane, $\gamma = 1.306$.

SOLUTION $R(\text{methane}) = \dfrac{R_u}{M} = \dfrac{8.314 \times 10^3}{16} = 0.52 \text{ kJ/kg K}$

Mass of methane initially present in the tank is given by

$$m_0 = \frac{P_0 V}{RT_0}$$

Mass of methane present in the tank at the end of the filling operation is given by

$$m_f = \frac{P_f V}{RT_f}$$

where V is the volume of the tank. For transient charging of an insulated tank, the First Law of Thermodynamics gives (see Eq. (4.40))

$$h_i(m_f - m_0) = m_f u_f - m_0 u_0$$

$$C_p T_i (m_f - m_0) = C_V T_f m_f - C_v T_0 m_0$$

$$\gamma T_i \left(\frac{P_f V}{RT_f} - \frac{P_0 V}{RT_0}\right) = T_f \left(\frac{P_f V}{RT_f}\right) - T_0 \left(\frac{P_0 V}{RT_0}\right)$$

or $$\gamma T_i \left(\frac{P_f}{T_f} - \frac{P_0}{T_0}\right) = P_f - P_0$$

$$1.306 \times 300 \left(\frac{20 \times 10^5}{T_f} - \frac{1 \times 10^5}{300}\right) = 20 \times 10^5 - 1 \times 10^5 \quad \text{or} \quad T_f = 385.9 \text{ K}$$

Mass of the methane which entered the tank is given by

$$m_f - m_0 = \frac{P_f V}{RT_f} - \frac{P_0 V}{RT_0} = \frac{V}{R}\left(\frac{P_f}{T_f} - \frac{P_0}{T_0}\right)$$

$$= \frac{1}{0.52 \times 10^3}\left(\frac{20 \times 10^5}{385.9} - \frac{1 \times 10^5}{300}\right) = 9.326 \text{ kg}$$

Discharging of a Cylinder Consider a cylinder containing a gas at high pressure. If the valve of the cylinder is opened, the gas inside the cylinder rushes out into the surroundings. Consider the cylinder as the control volume. The general expression for the First Law of Thermodynamics for a control volume is given by

$$\frac{dE}{dt} = \dot{m}_i(h_i + \mathrm{V}_i^2/2 + gz_i) - \dot{m}_e(h_e + \mathrm{V}_e^2/2 + gz_e) + \dot{Q} - \dot{W}_s \tag{4.26}$$

The control volume for discharging of a cylinder is shown in Fig. 4.16.

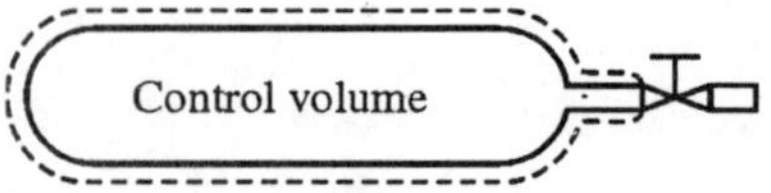

Fig. 4.16 Control volume for discharging of a cylinder.

While the gas escapes from the cylinder, the mass of gas in the CV as well as the energy of the CV changes. At any time the state of the gas leaving the cylinder will be the same as the state of the gas inside the cylinder. Since no mass is entering the cylinder, $\dot{m}_i = 0$. No shaft work is delivered by the CV and hence $\dot{W}_s = 0$. Assume that the cylinder is insulated. Then $\dot{Q} = 0$. Ignore the KE and PE terms. Then Eq. (4.26) reduces to

$$\int_0^t \frac{dE}{dt} = -\int_0^t \dot{m}_e h_e \tag{4.43}$$

To simplify the analysis, let us assume that h_e, the enthalpy of the gas at the exit is constant (if the variation in enthalpy of the exit stream is not much, this assumption is justified). Then Eq. (4.43) can be integrated to obtain

$$E_f - E_0 = -h_e(m_0 - m_f)$$

or $$m_0 u_0 - m_f u_f = (m_0 - m_f)h_e \tag{4.44}$$

Review Questions

1. What is the sign convention for work and heat?
2. State the First Law of Thermodynamics.
3. How is the work done by a system related to the heat interaction, if a system undergoes a cyclic change?
4. What is a cyclic process?

5. State the corollaries or the consequences of the First Law of Thermodynamics.
6. Prove that heat is a path function.
7. Suppose one adopts the sign convention that work done by the system is positive and heat transferred from the system to the surroundings is positive. State the First Law of Thermodynamics which is consistent with this sign convention.
8. Is it true that the energy of a system remains constant? If not, correct the statement.
9. What is a Perpetual Motion Machine of the First Kind?
10. A quartz wrist watch runs continuously and it does work in rotating the second, minute and hour hands without receiving heat input. Can it be called a PMMFK? If not, specify why.
11. Distinguish between non-flow and flow processes.
12. What is an isochoric process?
13. Name the process for which the change in the internal energy of the system is equal to the heat interaction.
14. Name the process for which heat interaction is equal to the change in the enthalpy of the system.
15. Name the process for which heat interaction is equal to a property change.
16. What will be the change in the internal energy if a fluid contained in a jar is whipped for 60 seconds using a 400 W mixer/grinder?
17. Is it possible to increase the internal energy of a fluid contained in an insulated vessel? Explain.
18. Give some examples of constant pressure processes.
19. How is the heat interaction related to work done, if an ideal gas undergoes an isothermal process?
20. During a given process is it possible to convert all the heat supplied to a system into work? If so, give an example.
21. What is an adiabatic process?
22. What is the relation between P and v for an ideal gas, if it follows a reversible adiabatic path?
23. What is a polytropic process?
24. Name the polytropic processes Pv^n = constant, for $n = 0$ and $n = \infty$.
25. What is the work done by a system if it undergoes a change of state from P_1, V_1 to P_2, V_2 by following a reversible adiabatic path? Will the work done be the same if the path is adiabatic but irreversible? Explain.
26. Starting from the First Law of Thermodynamics given by $\Delta E = Q - W$ for a closed system, derive the first law expression for a control volume.
27. What is a steady-state, steady-flow process?
28. What is a throttling process?
29. Which thermodynamic property of a fluid remains unaltered during throttling?
30. Define the Joule–Thomson coefficient.
31. What is inversion point and what is inversion curve?
32. What is the value of the Joule–Thomson coefficient for an ideal gas?
33. Will there be any change in temperature if an ideal gas is throttled from high pressure to low pressure?

Problems

4.1 A system is taken from the initial state i to the final state f along the path i-a-f. During this process 50 kJ of work was done by the system while it absorbed 75 kJ heat. If the system is made to undergo the same change of state by a different path i-b-f in which it absorbs 105 kJ heat, how much work is done by the system? If the system is restored to its initial state i from state f through an adiabatic path, what is the work done by the system?

4.2 A system is changed from state 1 to state 2 by a path A. During this process 10 kJ of work was done on the system and it rejected 25 kJ heat. It is desired to restore the system to state 1 by following an adiabatic process. Calculate the heat and work interactions for this process.

4.3 A system is taken through four processes and the final state of the system is identical with the initial state. The measured work and heat interactions during the processes are given below.

Process	W (kJ)	Q (kJ)	ΔU (kJ)
1-2	200	100	–
2-3	100	–	100
3-4	–150	–	–
4-1	50	0	–

Complete the missing measurements and determine the net heat interaction.

4.4 An ideal gas is heated at constant volume until its temperature is doubled and then cooled at constant pressure until it returns to the original temperature. Finally, the gas is allowed to expand isothermally till it returns to the original volume. Determine the net heat interaction.

4.5 An insulated piston-cylinder assembly contains 0.1 kg air (ideal gas, $\gamma = 1.4$) at 1 bar and 300 K. An electrical resistor placed inside the gas space is connected to a battery through a switch. The switch is closed, allowing the current to pass through the resistor till the volume of the gas is doubled while the pressure is held constant. Calculate the work done by the gas and the energy dissipated by the electrical resistor.

4.6 A piston-cylinder assembly contains 0.1 mol of air (ideal gas, $\gamma = 1.4$) at 1 bar and 300 K. If it is compressed reversibly and adiabatically to 10 bar pressure, determine the final temperature and volume of the air.

4.7 An ideal gas ($\gamma = 1.67$) undergoes reversible and adiabatic expansion from 5 bar and 500 K till its temperature is reduced to 400 K. Determine the final pressure.

4.8 It is known that reversible adiabatic expansion of a gas reduces its temperature. Calculate the final temperature if air ($\gamma = 1.4$) at 2 bar and 300 K is reversibly and adiabatically expanded to 1 bar.

4.9 A piston-cylinder assembly contains 0.05 kg of Helium at 1 bar and 300 K. If the Helium is compressed till its final volume is one-eighth of the initial volume by following the path $PV^{1.4}$ = constant, determine the work and heat interactions. For Helium, $C_p = 5.19$ kJ/kg K. Assume that Helium is an ideal gas.

4.10 An insulated turbine in a steam power plant, operating in steady-state, is fed with steam with an enthalpy of 3230 kJ/kg. The steam enters the turbine with a velocity of 10 m/s at an elevation of 5 m above the ground level. After expansion, the low pressure steam with an enthalpy of 2580 kJ/kg leaves the turbine with a velocity of 50 m/s at an elevation of 3 m above the ground level. If the power output of the turbine is 400 MW, estimate the mass flow rate of steam through the turbine.

4.11 Calculate the percent error, if one ignores the change in the KE of steam for Problem 4.10.

4.12 An air compressor compresses air (ideal gas, $\gamma = 1.4$) from 1 bar and 300 K to 10 bar and 600 K. If the flow rate of air through the compressor is 0.1 kg/s, calculate the power consumption of the compressor, assuming that the compressor is adiabatic.

4.13 In a steam power plant, liquid water ($\rho = 1000$ kg/m^3) is to be pumped into a boiler which is at 30 bar. Assuming that the pump operates under adiabatic conditions and receives liquid water at 1 bar, calculate the power required to operate the pump, if the flow rate of liquid through the pump is 200 kg/s. Assume liquid water to be an incompressible substance.

4.14 Air (ideal gas, $\gamma = 1.4$) expands reversibly and adiabatically through a nozzle from 2 bar and 27°C to a pressure of 1 bar. Determine the exit velocity of air, if air enters the nozzle with negligible velocity.

4.15 Steam at 1.5 bar and 120°C (h = 2711.4 kJ/kg), enters an adiabatic nozzle with negligible velocity and leaves at 1 bar with an enthalpy of 2676 kJ/kg. Calculate the exit velocity of steam.

4.16 An evacuated and insulated tank of 1 m^3 capacity is connected to a line carrying air (ideal gas, $\gamma = 1.4$) at 30 bar and 300 K. The air enters the tank till the air pressure in the tank is identical with the air pressure in the filling line. Estimate the final temperature of air in the tank and the mass of air which enters the tank.

4.17 An insulated tank of 10 m^3 capacity contains air (ideal gas, $\gamma = 1.4$) at 1 bar and 27°C. This tank is connected to a line in which air at 50 bar and 300 K is flowing and the tank is filled with air to a pressure of 50 bar. Determine the final temperature and mass of air in the tank.

5 SECOND LAW OF THERMODYNAMICS

The First Law of Thermodynamics was discussed in chapter 4 and it was applied to non-flow and flow processes to obtain several useful results. The First Law of Thermodynamics is essentially a law of conservation of energy and does not specify the direction or feasibility of a process. In this chapter, the Second Law of Thermodynamics is presented.

5.1 Limitations of the First Law of Thermodynamics

Consider the example of water flowing downhill. While the water is on the hilltop, it possesses PE. As it flows down, its PE decreases and KE increases. The gain in the KE is exactly equal to the loss in PE. If the water flows down further and is about to touch the ground, it possesses no PE and all the PE of the water is converted into KE. Once the water reaches the ground and comes to rest, it does not possess any KE and its internal energy increases. The increase in the internal energy will be exactly equal to the decrease in the KE.

Now, imagine a reversal of the above process. That is, water which is at the foot of the hill, on its own gaining KE and flowing up while its internal energy decreases. For this process also, the First Law of Thermodynamics will be satisfied and the gain in the KE is equal to the loss in the internal energy. However, we never observe such a thing happening on its own. That is, the reversal of a spontaneous process on its own is never observed in nature. The First Law of Thermodynamics does not deny the possibility of a spontaneous process proceeding in the reverse direction on its own. This example shows that there exists a directional law which determines the direction in which a spontaneous process takes place. This directional law is called the Second Law of Thermodynamics.

The Second Law of Thermodynamics determines the feasibility of a process and if it is feasible, it dictates the limit or extent to which a process can proceed. Before we discuss the Second Law of Thermodynamics, let us understand the meaning of a few terms like thermal reservoir, heat engine, refrigerator, heat pump, etc.

5.2 Heat Engine, Refrigerator and Heat Pump

A *thermal reservoir* is a large system from which a quantity of energy can be absorbed or added as heat without changing its temperature. The atmosphere and the sea are examples of thermal reservoirs. Addition to or extraction of a certain quantity of energy as heat from the sea does not change its temperature.

A thermal reservoir or a body at high temperature from which energy is absorbed as heat by a heat engine is called a *source*. A *sink* is a thermal reservoir at low temperature to which a heat engine rejects energy as heat.

A *heat engine* is a cyclically operating device which absorbs energy as heat from a high temperature reservoir, converts part of the energy into work and rejects the rest of the energy as heat to a thermal reservoir at low temperature. The *working fluid* or the working medium is a substance which absorbs energy as heat from a source, and rejects energy as heat to a sink while undergoing a cyclic change in a heat engine. A thermal power plant shown in Fig. 5.1 is an example of a heat engine.

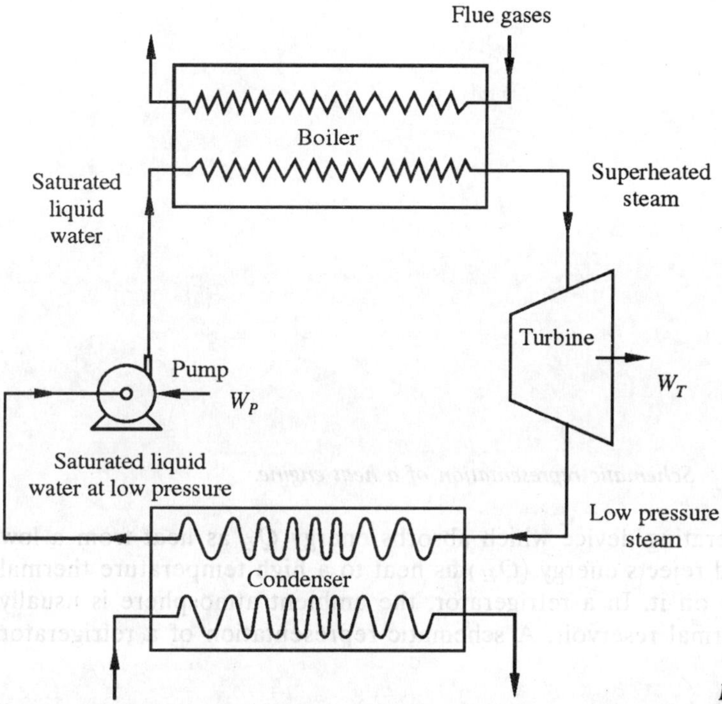

Fig. 5.1 *Schematic representation of a thermal power plant.*

In a thermal power plant, the working medium (water) at high pressure is fed to a boiler where it receives energy (Q_1) as heat from the hot gases obtained by burning coal or some other fuel. The water is converted into superheated steam in the boiler. The superheated steam leaving the boiler enters a turbine and expands to a low pressure. The turbine delivers the (shaft work) work W_T. The low pressure steam then passes through a condenser where it rejects (Q_2) as heat to the cooling water at low temperature and leaves the condenser as saturated liquid water at low pressure. Then the low pressure water is pumped to the boiler and work (W_p) is done on the pump. Thus, the working medium undergoes a cyclic change.

The index of performance of a heat engine is expressed by the *thermal efficiency* η which is defined as the ratio of the net work done to the energy absorbed as heat. A heat engine is schematically represented in Fig. 5.2.

Consider the working medium which is undergoing a cyclic change in a heat engine as the system (shown by the dotted boundary in Fig. 5.2) and apply the First Law of Thermodynamics to obtain

$$W = Q_1 - Q_2$$

Then, the thermal efficiency η of the heat engine is given by

$$\eta = \frac{W_{\text{net}}}{Q_1} = \frac{W}{Q_1} = \frac{Q_1 - Q_2}{Q_1} \tag{5.1}$$

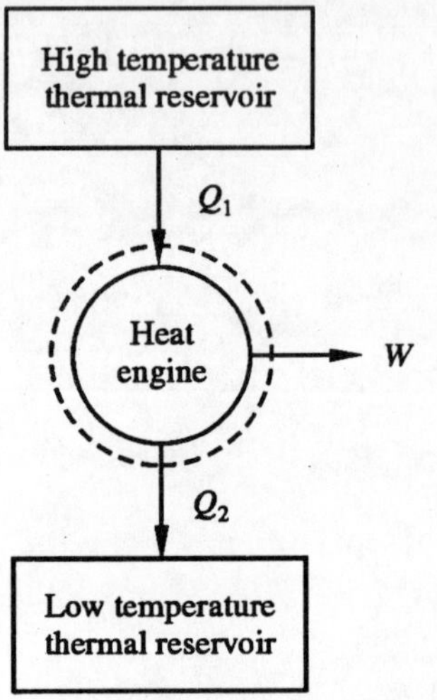

Fig. 5.2 Schematic representation of a heat engine.

A *refrigerator* is a cyclically operating device which absorbs energy Q_L as heat from a low temperature thermal reservoir and rejects energy (Q_H) as heat to a high temperature thermal reservoir when work (W) is done on it. In a refrigerator, the ambient atmosphere is usually used as the high temperature thermal reservoir. A schematic representation of a refrigerator is shown in Fig. 5.3(a).

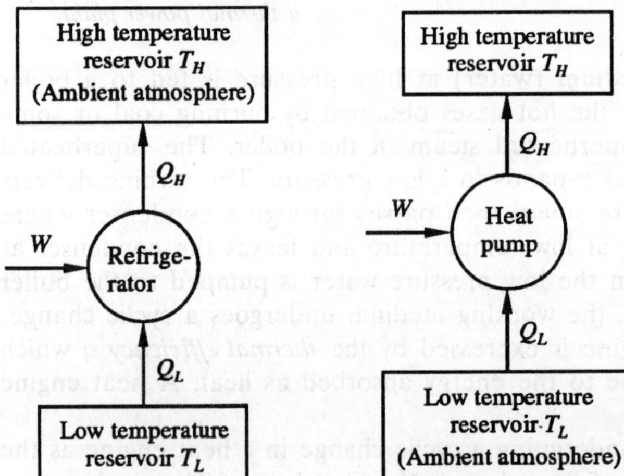

Fig. 5.3 Schematic representation of a (a) refrigerator. (b) heat pump.

A *heat pump* is a cyclically operating device which absorbs energy (Q_L) as heat from a low temperature thermal reservoir and rejects energy (Q_H) as heat to a high temperature thermal reservoir when work (W) is done on it. In a heat pump, the ambient atmosphere is usually used as a low temperature thermal reservoir. A heat pump is used in heating buildings in winter. A schematic representation of the heat pump is shown in Fig. 5.3(b). It can be observed that the refrigerator and heat pump are basically the same device, except that their objectives differ. In a refrigerator, the objective is to refrigerate and to maintain a body at low temperature by absorbing energy as heat from that body. On the other hand, if the objective of the device is to maintain a body at high temperature by rejecting energy as heat to it, it is called a heat pump.

The index of performance of the device—refrigerator and heat pump—is expressed in terms of the coefficient of performance (COP) which is defined as the ratio of the energy effect sought to the energy input required to accomplish the task. In a refrigerator, the energy effect sought is the extraction of heat Q_L from the low temperature reservoir and in the case of a heat pump, the energy effect sought is the rejection of heat Q_H to the high temperature thermal reservoir. Therefore, the coefficient of performance of a refrigerator $(\text{COP})_\text{R}$ is given by

$$(\text{COP})_\text{R} = \frac{Q_L}{W} = \frac{Q_L}{Q_H - Q_L} \tag{5.2}$$

The coefficient of performance of a heat pump $(\text{COP})_\text{HP}$ is given by

$$(\text{COP})_\text{HP} = \frac{Q_H}{W} = \frac{Q_H}{Q_H - Q_L} \tag{5.3}$$

It can be easily observed from Eqs. (5.2) and (5.3) that the $(\text{COP})_\text{R}$ and $(\text{COP})_\text{HP}$ are related to each other.

$$\frac{Q_L}{Q_H - Q_L} + 1 = \frac{Q_L + Q_H - Q_L}{Q_H - Q_L} = \frac{Q_H}{Q_H - Q_L}$$

or

$$(\text{COP})_\text{HP} = 1 + (\text{COP})_\text{R} \tag{5.4}$$

5.3 Second Law of Thermodynamics

Having understood the meaning of the terms thermal reservoir, heat engine, refrigerator, etc., we are now in a position to discuss the Second Law of Thermodynamics. There are several ways in which the Second Law of Thermodynamics can be stated. However, any statement of the Second Law of Thermodynamics, in essence, is to deny the possibility of a spontaneous process proceeding in the reverse direction on its own. Only two important statements of the Second Law of Thermodynamics, namely the Kelvin–Planck statement and the Clausius statement are presented below.

Kelvin–Planck Statement The Kelvin–Planck statement of the Second Law of Thermodynamics states that it is impossible to devise a cyclically operating device which produces no other effect than the extraction of heat from a single thermal reservoir and delivers an equivalent amount of work.

The Kelvin–Planck statement tells us that no heat engine can operate with only one thermal reservoir from which it absorbs energy and converts all the heat absorbed into work. A heat engine which operates with only one thermal reservoir is sometimes known as a 1-T engine and is shown in Fig. 5.4.

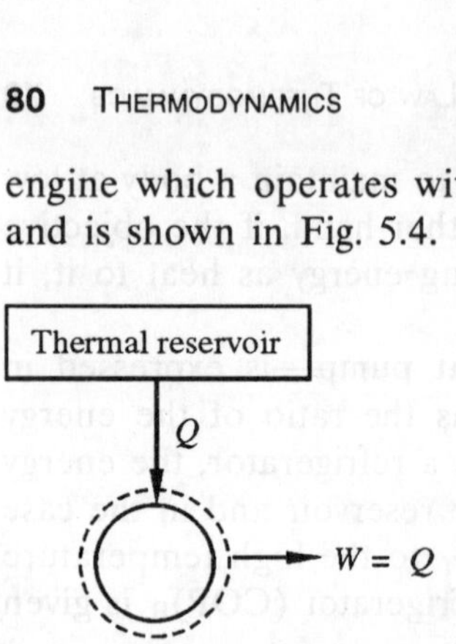

Fig. 5.4 Schematic representation of a 1-T engine.

For a 1-T engine, the thermal efficiency $\eta = (W/Q) = 1$. Thus the Kelvin–Planck statement tells us that no heat engine can have a thermal efficiency equal to unity.

Perpetual Motion Machine of the Second Kind A Perpetual motion machine of the second kind (PMMSK) is a cyclically operating device which absorbs energy as heat from a single thermal reservoir and delivers an equivalent amount of work. The Kelvin–Planck statement of the Second Law of Thermodynamics tells us that it is impossible to devise a PMMSK.

Clausius Statement The Clausius statement of the Second Law of Thermodynamics states that it is impossible to devise a self-acting cyclically operating device which transfers energy as heat from a low temperature body to a high temperature body.

The flow of heat from a high temperature body to a low temperature body is a spontaneous process. The Clausius statement tells us that it is impossible to transfer heat from a low temperature body to a high temperature body unless the device used for this purpose is externally aided. In other words, the Clausius statement denies the possibility of the spontaneous flow of heat from a low temperature body to a high temperature body on its own.

5.4 Equivalence of the Kelvin–Planck and the Clausius Statements

The Kelvin–Planck and Clausius statements of the Second Law of Thermodynamics appear to be altogether different. However, the two statements are equivalent. The equivalence of these two statements can be proved by showing that the violation of one statement leads to a violation of the other statement.

To prove that violation of the Kelvin–Planck statement leads to a violation of the Clausius statement, let us assume that the Kelvin–Planck statement is incorrect. That is, it is possible to construct a cyclically working device 1, which absorbs energy (Q_1) as heat from a thermal reservoir at T_H and performs an equivalent amount of work W ($W = Q_1$). Let us consider a cyclically operating device 2, which absorbs energy (Q_L) as heat from a low temperature thermal reservoir at T_L and rejects energy Q_H ($Q_H = Q_L + W$) when work W is done on it. Such a device does not violate the Clausius statement. The two devices are shown in Fig. 5.5(a). Now, we can combine the two devices 1 and 2 as shown in Fig. 5.5(b).

The device 1 receives heat Q_1 from a reservoir at T_H while device 2 rejects heat Q_H ($Q_H = Q_L + W = Q_L + Q_1$) to the reservoir at T_H. Therefore, part of the energy rejected by device 2 can be supplied to device 1 and the remaining energy ($Q_H - Q_1 = Q_L$) can be rejected to the reservoir at T_H as shown in Fig. 5.5(b). The combined device (enclosed by the dotted boundary) transfers heat Q_L from the low temperature reservoir at T_L to the high temperature reservoir at T_H without receiving any aid from an external agent, which is in violation of the Clausius statement. Therefore, it can be concluded that violation of the Kelvin–Planck statement leads to violation of the Clausius statement.

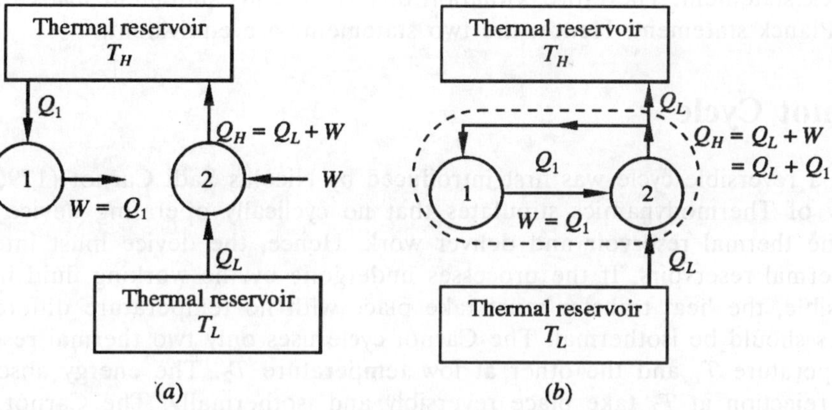

Fig. 5.5 Schematic representation to show that violation of the Kelvin–Planck statement leads to a violation of the Clausius statement.

To prove that violation of the Clausius statement leads to violation of the Kelvin–Planck statement, let us assume that the Clausius statement is incorrect. That is, we can devise a self-acting cyclically working device 1, which can transfer heat Q from a low temperature thermal reservoir at T_L to a high temperature thermal reservoir at T_H. Consider another device 2 which absorbs heat Q_1 from a high temperature reservoir at T_H, does work W and rejects energy Q as heat to the low temperature reservoir at T_L. The two devices are shown in Fig. 5.6(a).

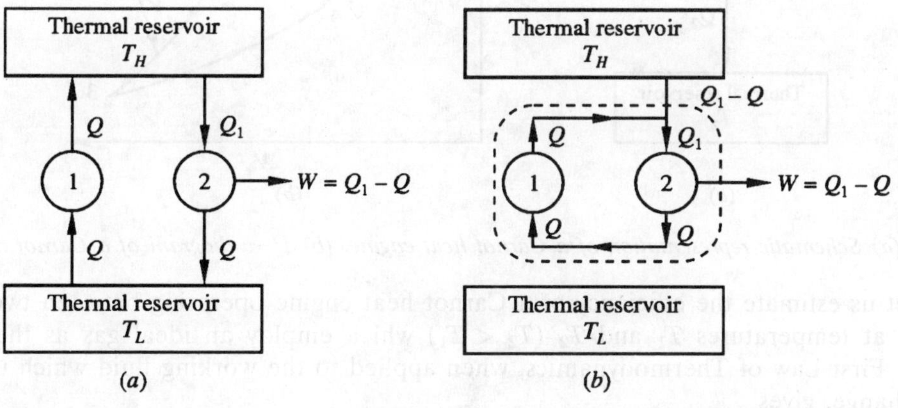

Fig. 5.6 Schematic representation to prove that violation of the Clausius statement leads to violation of the Kelvin–Planck statement.

Now, the two devices can be combined as shown in Fig. 5.6(b), where the heat Q rejected by device 2 (at temperature T_L) is directly supplied to the device 1 which rejects heat Q at temperature T_H. This energy Q as heat can be supplied to device 2, and the device 2 absorbs the remaining energy $(Q_1 - Q)$ from the reservoir at T_H as shown in Fig. 5.6(b). This combined unit (shown in the figure by a dotted enclosure) receives energy $(Q_1 - Q)$ as heat from a single thermal reservoir and delivers work W $(W = Q_1 - Q)$ in violation of the Kelvin–Planck statement. Therefore, violation of the Clausius statement leads to violation of the Kelvin–Planck statement. Hence, the two statements are equivalent.

5.5 Carnot Cycle

The idea of a reversible cycle was first introduced by Nicolas Sadi Carnot (1796–1832). The Second Law of Thermodynamics stipulates that no cyclically operating device can interact with only one thermal reservoir and deliver work. Hence, the device must interact with at least two thermal reservoirs. If the processes undergone by the working fluid in a cycle are to be reversible, the heat transfer must take place with no temperature difference, that is, the processes should be isothermal. The Carnot cycle uses only two thermal reservoirs—one at high temperature T_1 and the other at low temperature T_2. The energy absorption at T_1 and energy rejection at T_2 take place reversibly and isothermally. The Carnot cycle shown in Fig. 5.7 consists of a reversible isothermal expansion from state 1 to state 2, reversible adiabatic expansion from state 2 to state 3, a reversible isothermal compression from state 3 to state 4 followed by a reversible adiabatic compression to state 1.

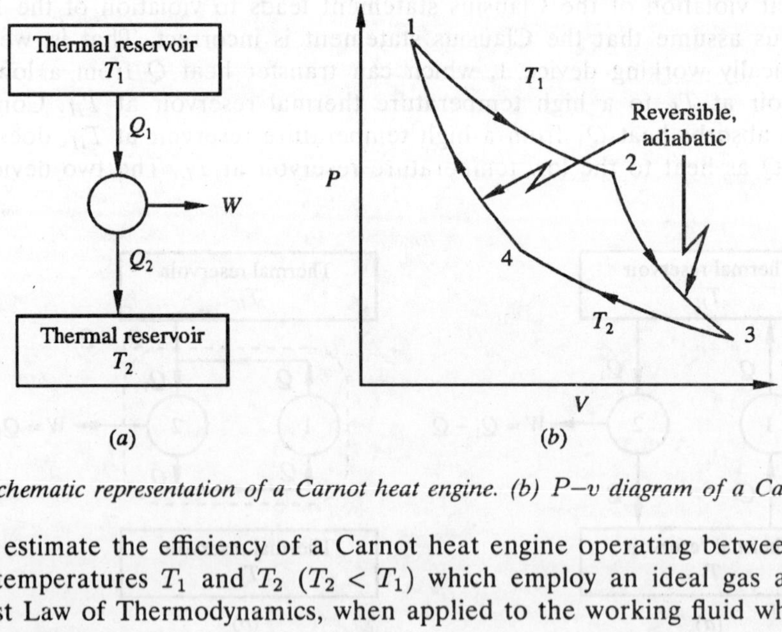

Fig. 5.7 (a) Schematic representation of a Carnot heat engine. (b) P–v diagram of a Carnot cycle.

Now let us estimate the efficiency of a Carnot heat engine operating between two thermal reservoirs at temperatures T_1 and T_2 $(T_2 < T_1)$ which employ an ideal gas as the working fluid. The First Law of Thermodynamics, when applied to the working fluid which undergoes a cyclic change, gives

$$\oint dW = \oint dQ \tag{5.5}$$

That is, the net work done is equal to the net heat interaction. The thermal efficiency η is given by

$$\eta = \frac{\text{Net work done}}{\text{Energy absorbed as heat}} \tag{5.6}$$

Since processes 2-3 and 4-1 (see Fig. 5.7(b)) are adiabatic, the heat interaction for these processes is equal to zero. The process 1-2 is reversible, isothermal and for an ideal gas $u = u(T)$ only. Therefore,

$$du = 0 = dQ - dW$$

or

$$Q_{1-2} = \int_1^2 P dv = \int_1^2 RT_1 \frac{dv}{v} = RT_1 \ln(v_2/v_1) \tag{5.7}$$

Similarly, for the process 3-4 which is also reversible and isothermal, we get

$$Q_{3-4} = \int_3^4 P dv = \int_3^4 RT_2 \frac{dv}{v} = RT_2 \ln(v_4/v_3) \tag{5.8}$$

Net heat interaction = Net work done = $RT_1 \ln(v_2/v_1) + RT_2 \ln(v_4/v_3)$

$$= RT_1 \ln(v_2/v_1) - RT_2 \ln(v_3/v_4) \tag{5.9}$$

The processes 2-3 and 4-1 are reversible, adiabatic and hence

$$T_1 v_2^{\gamma-1} = T_2 v_3^{\gamma-1} \quad \text{or} \quad \frac{v_2}{v_3} = \left(\frac{T_2}{T_1}\right)^{1/\gamma-1} \tag{5.10}$$

and

$$T_2 v_4^{\gamma-1} = T_1 v_1^{\gamma-1} \quad \text{or} \quad \frac{v_1}{v_4} = \left(\frac{T_2}{T_1}\right)^{1/\gamma-1} \tag{5.11}$$

From Eqs. (5.10) and (5.11), we get

$$\frac{v_2}{v_3} = \frac{v_1}{v_4} \quad \text{or} \quad \frac{v_2}{v_1} = \frac{v_3}{v_4} \tag{5.12}$$

Substituting Eq. (5.9) in Eq. (5.6), we get

$$\eta = \frac{RT_1 \ln(v_2/v_1) - RT_2 \ln(v_3/v_4)}{RT_1 \ln(v_2/v_1)} \tag{5.13}$$

From Eqs. (5.12) and (5.13), we get

$$\eta = \frac{T_1 - T_2}{T_1} = 1 - \frac{T_2}{T_1} \tag{5.14}$$

5.6 Carnot Theorems

The consequences of the Second Law of Thermodynamics were proposed by Carnot and are called Carnot theorems. Carnot theorems are discussed below.

Carnot Theorem 1 No heat engine operating between the two given thermal reservoirs can be more efficient than a reversible heat engine operating between the same two thermal reservoirs.

To prove this theorem, let us assume that the theorem is incorrect. That is, it is possible for an engine I to have an efficiency greater than the efficiency of a reversible heat engine R when both of them are operated between the same two thermal reservoirs. That is,

$$\eta_I > \eta_R \tag{5.15}$$

Let the engines I and R absorb the same quantity of energy Q_1 as heat from a thermal reservoir at T_1. Let Q and Q_2 represent the energy rejected as heat by the engines I and R, respectively, to the thermal reservoir at T_2. Then the work done by the engines is given by

$$W_I = Q_1 - Q \tag{5.16}$$

$$W_R = Q_1 - Q_2 \tag{5.17}$$

and their efficiencies are

$$\eta_I = \frac{W_I}{Q_1} = \frac{Q_1 - Q}{Q_1} = 1 - \frac{Q}{Q_1} \tag{5.18}$$

$$\eta_R = \frac{W_R}{Q_1} = \frac{Q_1 - Q_2}{Q_1} = 1 - \frac{Q_2}{Q_1} \tag{5.19}$$

Since $\eta_I > \eta_R$, we have

$$1 - \frac{Q}{Q_1} > 1 - \frac{Q_2}{Q_1} \quad \text{or} \quad Q < Q_2 \tag{5.20}$$

Therefore

$$W_I (= Q_1 - Q) > W_R (= Q_1 - Q_2) \tag{5.21}$$

That is, for the same amount of energy input Q_1, the engine I delivers more work than the engine R. Since the engine R is reversible it can be made to execute the cycle in the reverse order. Then, it will absorb energy Q_2 from the reservoir at T_2 and reject energy Q_1 to the reservoir at T_1 when work W_R is done on it. Now let us combine the engines I and R as shown in Fig. 5.8.

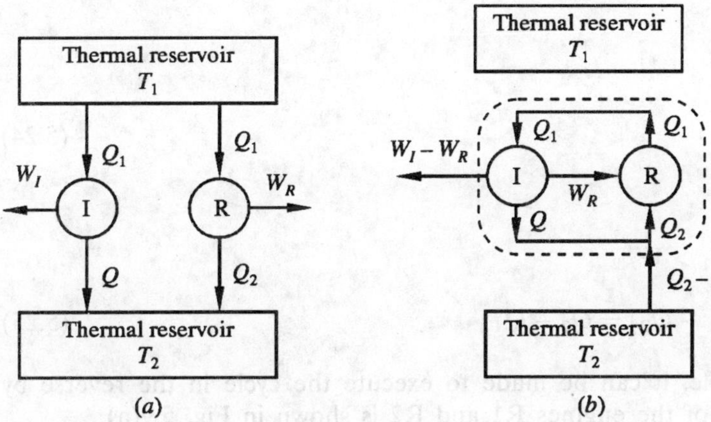

Fig. 5.8 *Schematic representation to prove Carnot theorem 1.*

The network delivered by the combined device (engine I and R put together) is given by

$$W_I - W_R = Q_1 - Q - (Q_1 - Q_2) = Q_2 - Q \tag{5.22}$$

The combined device absorbs energy $(Q_2 - Q)$ as heat from a single thermal reservoir and delivers an equivalent amount of work, which is in violation of the Second Law of Thermodynamics. Therefore, the assumption that $\eta_I > \eta_R$ is not correct. Hence,

$$\eta_R \geq \eta_I \tag{5.23}$$

Carnot Theorem 2 All reversible heat engines operating between the two given thermal reservoirs have the same efficiency. The efficiency of a reversible heat engine does not depend on the working fluid, but depends only on the temperatures of the reservoirs between which it operates.

To prove this Carnot theorem, let us consider two reversible heat engines R1 and R2, operating between the two given thermal reservoirs at temperatures T_1 and T_2. Let us assume that the efficiency (η_{R1}) of reversible heat engine R1 is greater than the efficiency (η_{R2}) of reversible heat engine R2.

Let Q_1 = energy absorbed as heat from the reservoir at T_1 by the engines R1 and R2, separately.

Q = energy rejected by reversible engine R1 to the reservoir at T_2.

Q_2 = energy rejected by reversible engine R2 to the reservoir at T_2

$W_{R1} = Q_1 - Q$ = work done by reversible engine R1

$W_{R2} = Q_1 - Q_2$ = work done by reversible engine R2

We have assumed that

$$\eta_{R1} > \eta_{R2}$$

or $$1 - \frac{Q}{Q_1} > 1 - \frac{Q_2}{Q_1}$$

or $$Q < Q_2 \tag{5.24}$$

Therefore,

$$Q_1 - Q > Q_1 - Q_2 \quad \text{or} \quad W_{R1} > W_{R2}$$

$$W_{R1} - W_{R2} = (Q_1 - Q) - (Q_1 - Q_2) = Q_2 - Q \tag{5.25}$$

Since the engine R2 is reversible, it can be made to execute the cycle in the reverse by supplying work W_{R2}. A schematic of the engines R1 and R2 is shown in Fig. 5.9(a).

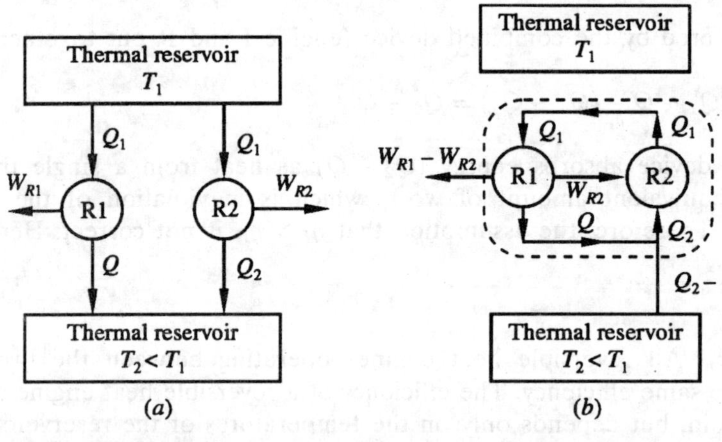

Fig. 5.9 Schematic representation of (a) reversible engines R1 and R2. (b) combined device which is a PMMSK.

Since $W_{R1} > W_{R2}$, the reversible engine R2 can be run as a heat pump by utilizing part of the work delivered by R1 as shown in Fig. 5.9(b). The combined device delivers work $W_{R1} - W_{R2} = Q_2 - Q$, by absorbing energy $(Q_2 - Q)$ from a single thermal reservoir. Hence the combined device is a PMMSK which is in violation of the Second Law of Thermodynamics. Hence our assumption that $\eta_{R1} > \eta_{R2}$ is incorrect. Therefore, we conclude that

$$\eta_{R2} \geq \eta_{R1} \tag{5.26}$$

Now let us assume that $\eta_{R2} > \eta_{R1}$ and by adopting similar arguments as above, we can reach the conclusion that

$$\eta_{R1} \geq \eta_{R2} \tag{5.27}$$

Based on Eqs. (5.26) and (5.27), it can be concluded that

$$\eta_{R1} = \eta_{R2} \tag{5.28}$$

That is, all reversible engines operating between the two given thermal reservoirs have the same efficiency.

The Carnot theorem 2 further states that the efficiency of a reversible heat engine does not depend on the working fluid, but depends only on the temperatures of the reservoirs between which it operates. To prove this, consider two reversible heat engines $R1$ and $R2$ which use different working fluids. By following the arguments presented above one can show that the efficiency of a reversible heat engine is independent of the working fluid and depends only on the temperatures of the reservoirs between which it operates.

5.7 Thermodynamic Temperature Scale

The Zeroth Law of Thermodynamics and temperature scales were discussed in chapter 3. It was stated that the relationship between thermometric property and temperature is not linear and hence different thermometers read different values at a given temperature except at the fixed points which were used in calibrating them. Therefore, it is desirable to define a temperature scale which does not depend on the thermometric property of a substance.

We have learnt from the Carnot theorem that the efficiency of a Carnot engine does not depend on the working fluid but depends only on the temperature of the reservoirs between which it operates. This fact can be utilized to establish a temperature scale. Since the Carnot theorems are consequences of the Second Law of Thermodynamics, the temperature scale so established is called the *thermodynamic temperature scale or absolute temperature scale.*

To establish the thermodynamic temperature scale, consider the operation of three reversible engines 1, 2 and 3. The engine 1 absorbs energy Q_1 as heat from the reservoir at θ_1, do work W_1 and rejects energy Q_2 as heat to the reservoir at θ_2. Let the engine 2 absorb energy Q_2 as heat from the reservoir at θ_2, do work W_2 and reject energy Q_3 as heat to the reservoir at θ_3. The third reversible engine 3 absorbs energy Q_1 as heat from the reservoir at θ_1, does work W_3 and rejects energy Q_3 as heat to the reservoir at θ_3, where θ denotes the temperature as measured on the thermodynamic temperature scale. A schematic arrangement of the three reversible engines is shown in Fig. 5.10.

$$\eta_1 = \frac{W_1}{Q_1} = 1 - \frac{Q_2}{Q_1} = f(\theta_1, \theta_2)$$

or

$$\frac{Q_1}{Q_2} = F(\theta_1, \theta_2) \tag{5.29}$$

$$\eta_2 = 1 - \frac{Q_3}{Q_2} = f(\theta_2, \theta_3)$$

or

$$\frac{Q_2}{Q_3} = F(\theta_2, \theta_3) \tag{5.30}$$

$$\eta_3 = 1 - \frac{Q_3}{Q_1} = f(\theta_1, \theta_3)$$

or

$$\frac{Q_1}{Q_3} = F(\theta_1, \theta_3) \tag{5.31}$$

where f and F represent two different functions.

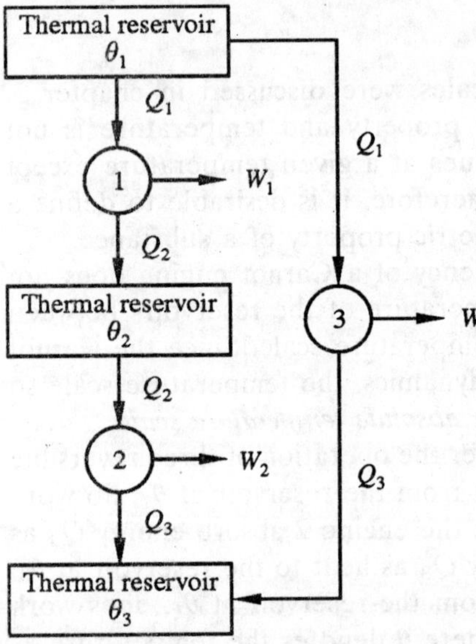

Fig. 5.10 *Schematic arrangement of three reversible engines to establish the thermodynamic temperature scale.*

Then $\quad \dfrac{Q_1}{Q_2} = \dfrac{Q_1/Q_3}{Q_2/Q_3}$

or

$$F(\theta_1, \theta_2) = \frac{F(\theta_1, \theta_3)}{F(\theta_2, \theta_3)} \tag{5.32}$$

In Eq. (5.32), θ_3 does not appear on the left-hand side. Therefore, on the right-hand side, the dependence of θ_3 should cancel out. This is possible if the function F can be written as

$$F(\theta_1, \theta_2) = \phi(\theta_1)\psi(\theta_2) \tag{5.33}$$

Then Eq. (5.32) can be rewritten as

$$\phi(\theta_1)\psi(\theta_2) = \frac{\phi(\theta_1)\psi(\theta_3)}{\phi(\theta_2)\psi(\theta_3)} = \frac{\phi(\theta_1)}{\phi(\theta_2)} \tag{5.34}$$

Therefore,

$$\psi(\theta_2) = \frac{1}{\phi(\theta_2)} \tag{5.35}$$

Hence

$$\frac{Q_1}{Q_2} = F(\theta_1, \theta_2) = \frac{\phi(\theta_1)}{\phi(\theta_2)} \tag{5.36}$$

Now, the temperature scale can be defined by choosing the function ϕ. Let us choose the simplest choice

$$\phi(\theta) = \theta \tag{5.37}$$

Then

$$\frac{Q_1}{Q_2} = \frac{\theta_1}{\theta_2} \tag{5.38}$$

That is, the ratio of the energy absorbed to the energy rejected as heat by a reversible engine is equal to the ratio of the temperatures of the source and the sink. The Eq. (5.38) can be used to determine the temperature of any reservoir by operating a reversible engine between that reservoir and another easily reproducible thermal reservoir and by measuring the heat interactions. The temperature of the easily reproducible thermal reservoir can be arbitrarily assigned a numerical value. The triple point of water (273.16 K) can be chosen as the easily reproducible thermal reservoir.

The efficiency of a Carnot engine operating between two thermal reservoirs, the temperatures of which are measured on the thermodynamic temperature scale, is given by

$$\eta = 1 - \frac{Q_2}{Q_1} = 1 - \frac{\theta_2}{\theta_1} \quad \text{or} \quad \frac{Q_2}{Q_1} = \frac{\theta_2}{\theta_1} \tag{5.39}$$

The efficiency of a Carnot engine, using an ideal gas as the working medium and the temperatures as measured on the ideal gas temperature scale is, given by

$$\eta = 1 - \frac{Q_2}{Q_1} = 1 - \frac{T_2}{T_1} \quad \text{or} \quad \frac{Q_2}{Q_1} = \frac{T_2}{T_1} \tag{5.40}$$

From Eqs. (5.39) and (5.40), we find that

$$\frac{\theta_2}{\theta_1} = \frac{T_2}{T_1} \tag{5.41}$$

Since the numerical value 273.16 K is assigned to the triple point of water in the thermodynamic temperature scale as well as in the ideal gas temperature scale, these two scales are equivalent.

Therefore, the thermodynamic temperature is numerically equal to the temperature measured by an ideal gas thermometer. Hence, hereafter, we use the symbol T only to represent the temperature.

Equation (5.40) which expresses the relation between the ratios Q_2/Q_1 and T_2/T_1 can be used to express the efficiency of a Carnot heat engine and the COP of a refrigerator/heat pump in terms of the absolute temperatures of the reservoirs between which they are operating.

$$\eta = 1 - \frac{Q_2}{Q_1} = 1 - \frac{T_2}{T_1}$$

$$(\text{COP})_\text{R} = \frac{Q_L}{Q_H - Q_L} = \frac{T_L}{T_H - T_L} \tag{5.42}$$

$$(\text{COP})_\text{HP} = \frac{Q_H}{Q_H - Q_L} = \frac{T_H}{T_H - T_L} \tag{5.43}$$

EXAMPLE 5.1 Suppose a huge body is at 127°C while the ambient atmosphere is at 27°C. It is proposed to run a heat engine which can absorb energy as heat from the body at 127°C and with the ambient atmosphere as the sink. Determine the maximum possible efficiency of the heat engine.

SOLUTION The efficiency will be maximum for a reversible heat engine. We also know that all reversible heat engines will have the same efficiency when operated between a given set of thermal reservoirs. For a Carnot heat engine, the efficiency is given by

$$\eta = 1 - \frac{T_2}{T_1} = 1 - \frac{273 + 27}{273 + 127} = 1 - \frac{300}{400} = 0.25$$

Therefore, the maximum possible efficiency = 0.25

EXAMPLE 5.2 An inventor claims to have developed a heat engine which delivers a power output of 1 kW while absorbing energy at the rate of 1.2 kW from a source at 927°C when the ambient atmosphere is at 27°C. Would you support this claim?

SOLUTION The maximum possible efficiency is given by

$$\eta = 1 - \frac{T_2}{T_1} = 1 - \frac{300}{1200} = 0.75$$

The efficiency claimed by the inventor

$$= \frac{\dot{W}}{\dot{Q}_1} = \frac{1 \times 10^3}{1.2 \times 10^3} = 0.833$$

As the efficiency claimed by the inventor is greater than the theoretically possible maximum efficiency, the claim is not justified.

EXAMPLE 5.3 It is proposed to devise a refrigeration plant to convert 100 tonnes (1 tonne = 1000 kg) of water at 0°C into ice at 0°C in a day in summer while the ambient atmosphere is at 37°C. The latent heat of fusion of water at 0°C is 334.92 kJ/kg. Estimate the minimum power required to operate the refrigeration plant.

SOLUTION The power required will be minimum when the refrigerator is reversible. The COP of a refrigerator is given by

$$(\text{COP})_\text{R} = \frac{T_L}{T_H - T_L} = \frac{273}{310 - 273} = 7.378$$

Heat load on the refrigerator is given by

$$\dot{Q}_L = \frac{100 \times 1000 \times 334.92 \times 10^3}{24 \times 60 \times 60} = 387.639 \text{ kW}$$

$$(\text{COP})_\text{R} = \frac{\dot{Q}_L}{\dot{W}} \quad \text{or} \quad 7.378 = \frac{387.639 \times 10^3}{\dot{W}} \quad \text{or} \quad \dot{W} = 52.54 \text{ kW}$$

EXAMPLE 5.4 The normal boiling point of Helium is 4.2 K and the latent heat of vaporization of Helium is 20.83 kJ/kg. It is required to produce 5 kg/hour liquid Helium from Helium vapor at 4.2 K while the surrounding atmosphere is at 27°C. Calculate the COP of the refrigerating unit and the power required to operate the refrigeration unit.

SOLUTION $$(\text{COP})_\text{R} = \frac{T_L}{T_H - T_L} = \frac{4.2}{300.15 - 4.2} = 0.0142$$

Heat removal rate, $\dot{Q}_L = \dfrac{5 \times 20.83 \times 10^3}{60 \times 60} = 28.93 \text{ W}$

$$(\text{COP})_\text{R} = \frac{\dot{Q}_L}{\dot{W}} \quad \text{or} \quad 0.0142 = \frac{28.93}{\dot{W}} \quad \text{or} \quad \dot{W} = 2.04 \text{ kW}$$

EXAMPLE 5.5 It is proposed to use a heat engine, which absorbs energy as heat at 127°C with the surrounding atmosphere at 27°C as the sink, to operate a refrigerator to produce ice at 0°C from water at 0°C. The refrigerator rejects energy to the surrounding atmosphere. Calculate (a) the ratio of the energy removed from the cold body by the refrigerator to the energy absorbed by the heat engine. (b) the ratio of the energy rejected to the surrounding atmosphere to the energy absorbed by the engine.

SOLUTION A schematic arrangement of the heat engine and refrigerator is shown in Fig. 5.11.

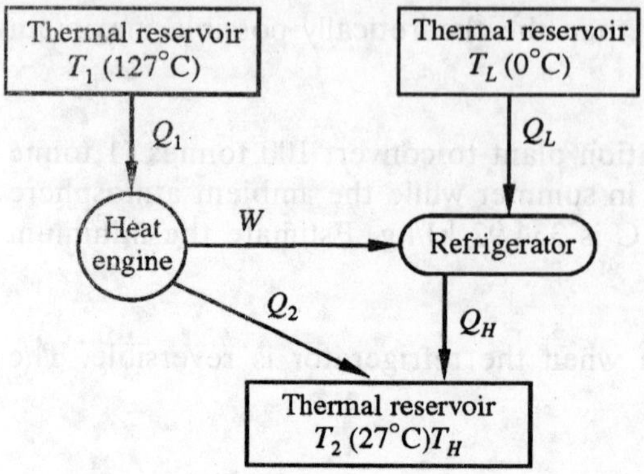

Fig. 5.11 *Schematic arrangement of heat engine driving a refrigerator.*

(a) Thermal efficiency of the heat engine is given by

$$\eta = 1 - \frac{300}{400} = 0.25 = \frac{W}{Q_1}$$

or

$$W = 0.25Q_1 \qquad \text{(A)}$$

The COP of the refrigerator is given by

$$(\text{COP})_R = \frac{273}{300 - 273} = 10.11 = \frac{Q_L}{W}$$

or

$$W = \frac{Q_L}{10.11} \qquad \text{(B)}$$

From Eqs. (A) and (B), we get

$$W = 0.25Q_1 = \frac{Q_L}{10.11} \quad \text{or} \quad \frac{Q_L}{Q_1} = 0.25 \times 10.11 = 2.528$$

(b) $$\eta = 1 - \frac{Q_2}{Q_1} = 1 - \frac{T_2}{T_1} \quad \text{or} \quad \frac{Q_2}{Q_1} = \frac{T_2}{T_1} = \frac{300}{400}$$

$$(\text{COP})_R = \frac{Q_L}{Q_H - Q_L} = \frac{T_L}{T_H - T_L} \quad \text{or} \quad \frac{Q_H}{Q_L} = \frac{T_H}{T_L} = \frac{300}{273} \quad \text{or} \quad Q_H = 1.099Q_L$$

Energy rejected to surrounding atmosphere = $Q_2 + Q_H = 0.75Q_1 + 1.099Q_L$

$$\text{Required ratio} = \frac{Q_2 + Q_H}{Q_1} = \frac{0.75Q_1 + 1.099Q_L}{Q_1}$$

$$= 0.75 + 1.099\frac{Q_L}{Q_1} = 0.75 + 1.099 \times 2.528 = 3.528$$

EXAMPLE 5.6 A reversible heat engine with an efficiency of 0.4 drives a reversible refrigerator with a COP of 5. If both the heat engine and the refrigerator reject energy to the surrounding atmosphere at 27°C, determine the temperatures of the reservoir from which the engine absorbs energy and the reservoir from which the refrigerator extracts energy.

SOLUTION $\eta = 1 - \dfrac{T_2}{T_1} = 1 - \dfrac{300}{T_1} = 0.4 \quad \text{or} \quad T_1 = 500 \text{ K}$

Therefore, the temperature of the reservoir from which the engine absorbs energy = 500 K

$$(\text{COP})_R = \frac{T_L}{T_H - T_L} = \frac{T_L}{300 - T_L} = 5 \quad \text{or} \quad T_L = 250 \text{ K}$$

The temperature of the reservoir from which the refrigerator extracts energy = 250 K

5.8 Clausius Inequality

We have observed that for a reversible heat engine

$$\frac{Q_2}{Q_1} = \frac{T_2}{T_1} \tag{5.40}$$

or

$$\frac{Q_1}{T_1} - \frac{Q_2}{T_2} = 0 \tag{5.44}$$

where Q_1 is the energy absorbed from the source (a positive quantity) and Q_2 is the energy rejected to the sink (a negative quantity). If we adopt the sign convention for heat interaction, Eq. (5.44) can be rewritten as

$$\frac{Q_1}{T_1} + \frac{Q_2}{T_2} = 0 \quad \text{or} \quad \sum \frac{Q}{T} = 0 \qquad \text{for a reversible cycle} \tag{5.45}$$

We know that the efficiency of an irreversible engine is less than the efficiency of a reversible engine, when both of them are operating between the same set of thermal reservoirs at temperature T_1 and T_2. That is,

$$\eta_I = 1 - \frac{Q_2'}{Q_1} < \eta_R = 1 - \frac{Q_2}{Q_1} = 1 - \frac{T_2}{T_1} \quad \text{or} \quad \frac{Q_2'}{Q_1} > \frac{T_2}{T_1} \quad \text{or} \quad \frac{Q_1}{T_1} - \frac{Q_2'}{T_2} < 0$$

If we use the sign convention for heat interactions, the above equation can be rewritten as

$$\frac{Q_1}{T_1} + \frac{Q_2'}{T_2} < 0 \quad \text{or} \quad \sum \frac{Q}{T} < 0 \qquad \text{for an irreversible cycle} \tag{5.46}$$

where Q_2' is the energy rejected by the irreversible engine to the sink.

Equations (5.45) and (5.46) can be combined and expressed as

$$\sum \frac{Q}{T} \le 0 \tag{5.47}$$

where the equality sign holds good for reversible processes and the inequality sign is true for irreversible processes. Equation (5.47) can be expressed in terms of differential heat interactions dQ as

$$\oint \frac{dQ}{T} \le 0 \tag{5.48}$$

which is known as the Clausius inequality. The Clausius inequality itself is quite often used as the Second Law of Thermodynamics.

EXAMPLE 5.7 A reversible heat engine shown in Fig. 5.12 operates between three thermal reservoirs at 600 K, 400 K and 300 K. The engine receives 2400 kJ of energy as heat from the reservoir at 600 K and does 1500 kJ of work. Determine the heat interactions with the other two thermal reservoirs.

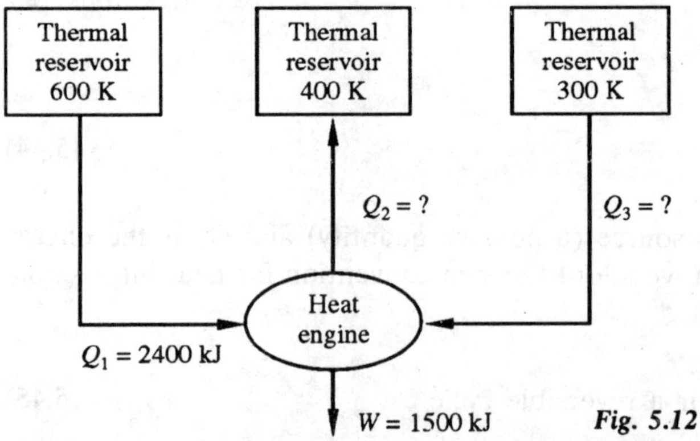

Fig. 5.12 *Schematic representation of a reversible heat engine operating with three thermal reservoirs.*

SOLUTION Let us assume that the heat engine rejects energy Q_2 and Q_3 as heat to the reservoirs at 400 K and 300 K, respectively. Then, the First Law of Thermodynamics gives

$$Q_1 - Q_2 - Q_3 = W$$

or $\quad 2400 - Q_2 - Q_3 = 1500$

or

$$Q_2 + Q_3 = 900 \text{ kJ} \tag{A}$$

The Second Law of Thermodynamics (in the form of Clausius inequality) for the reversible engine gives

$$\oint \frac{dQ}{T} = \frac{2400}{600} - \frac{Q_2}{400} - \frac{Q_3}{300} = 0$$

or

$$\frac{Q_2}{4} + \frac{Q_3}{3} = 400 \tag{B}$$

Solving Eqs. (A) and (B), we get

$$Q_2 = -1200 \text{ kJ} \quad \text{and} \quad Q_3 = 2100 \text{ kJ}$$

That is, the heat engine absorbs 1200 kJ of energy as heat from the reservoir at 400 K and rejects 2100 kJ of energy as heat to the reservoir at 300 K.

5.9 Entropy

The Second Law of Thermodynamics (in the form of Clausius inequality) can be used to analyze cyclic processes. The Second Law of Thermodynamics became a law of wide applicability when Clausius introduced the property entropy.

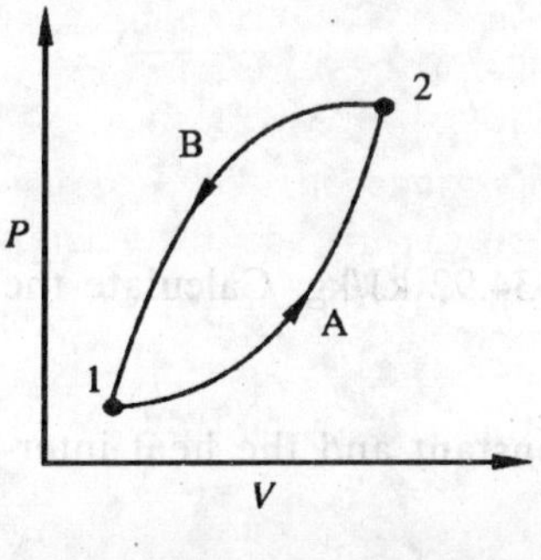

Fig. 5.13 A reversible cycle on P–v diagram to show that entropy is a property.

Consider a system which undergoes the cyclic change 1-A-2-B-1 as shown in Fig. 5.13. The processes 1-A-2 and 2-B-1 are reversible processes. Let us apply the Clausius inequality to the reversible cycle 1-A-2-B-1 and obtain

$$\oint \left(\frac{dQ}{T}\right)_{1A2B1} = 0 \quad \text{or} \quad \int \left(\frac{dQ}{T}\right)_{1A2} + \int \left(\frac{dQ}{T}\right)_{2B1} = 0$$

$$\int \left(\frac{dQ}{T}\right)_{1A2} = -\int \left(\frac{dQ}{T}\right)_{2B1} = \int \left(\frac{dQ}{T}\right)_{1B2} \tag{5.49}$$

Equation (5.49) shows that $\int (dQ/T)$ has the same value whether the path followed is 1-A-2 or 1-B-2. One can devise several reversible processes connecting the states 1 and 2 and find that $\int (dQ/T)$ has the same value for all these paths. Therefore, $(dQ/T)_R$ is an exact differential of some function, which is called entropy. The subscript R denotes that (dQ/T) is to be evaluated for a reversible path. Hence, we can say that there exists a property called entropy S, the change in which is given by

$$\Delta S = S_2 - S_1 = \int_1^2 dS = \int_1^2 \left(\frac{dQ}{T}\right)_R \tag{5.50}$$

The entropy is a property of a system and hence it is independent of the path followed by the system and it is a state function. The change in entropy of a system in reaching a given final state, from a given initial state is the same whether the system undergoes a reversible or an irreversible process. If a system undergoes a change of state by an irreversible process, one can devise an imaginary reversible process connecting the given initial and final states of the system and evaluate the entropy change for the imagined reversible process using Eq. (5.50).

EXAMPLE 5.8 Calculate the entropy change if 1 kg water at 30°C is heated to 80°C at 1 bar pressure. The specific heat of water is 4.2 kJ/kg K.

SOLUTION The initial and final states of water are given and the change in entropy depends on the initial and final states of the system only. If water is heated at constant pressure, the heat interaction is equal to the change in enthalpy. Hence

$$\Delta s = \int_1^2 \frac{dq}{T} = \int_1^2 \frac{dh}{T} = \int_1^2 \frac{C_p dT}{T} = C_p \ln \frac{T_2}{T_1}$$

$$= 4.2 \times 10^3 \ln \left(\frac{273 + 80}{273 + 30}\right) = 0.6415 \text{ kJ/kg K}$$

EXAMPLE 5.9 At 0°C, ice melts with a latent heat of fusion of 334.92 kJ/kg. Calculate the entropy change if 1 kg ice melts into water at 0°C.

SOLUTION During the phase change, the temperature remains constant and the heat interaction is equal to the latent heat of fusion. Therefore,

$$\Delta s = \int \frac{dq}{T} = \int \frac{dh}{T} = \frac{h_{sf}}{T} = \frac{334.92 \times 10^3}{273} = 1.2268 \text{ kJ/kg K}$$

EXAMPLE 5.10 A lump of steel of mass 30 kg at 427°C is dropped in 100 kg oil at 27°C. The specific heats of steel and oil are 0.5 kJ/kg K and 3.0 kJ/kg K, respectively. Calculate the entropy change of steel, oil and the universe.

SOLUTION Let T = final equilibrium temperature. The energy balance gives

$$(mC_p\Delta T)_{\text{Steel}} = (mC_p\Delta T)_{\text{oil}}$$

$$30 \times 0.5 \times (700 - T) = 100 \times 3(T - 300) \quad \text{or} \quad T = 319 \text{ K}$$

$$(\Delta S)_{\text{Steel}} = \int_1^2 \frac{dQ}{T} = \int_1^2 \frac{mC_p dT}{T} = \left(mC_p \ln \frac{T_2}{T_1}\right)_{\text{Steel}}$$

$$= 30 \times 0.5 \ln(319/700) = -11.7883 \text{ kJ/K}$$

$$(\Delta S)_{\text{oil}} = \left(mC_p \ln \frac{T_2}{T_1}\right)_{\text{oil}} = 100 \times 3 \ln \frac{319}{300} = 18.4226 \text{ kJ/K}$$

$$(\Delta S)_{\text{Universe}} = -11.7883 + 18.4226 = 6.6343 \text{ kJ/K}$$

5.10 Entropy Change of Ideal Gas

If an ideal gas undergoes a change of state from (P_1, v_1, T_1) to (P_2, v_2, T_2), the change in entropy can be calculated by devising a convenient reversible path connecting the given initial and final states. Then, the change in entropy can be calculated by making use of Eq. (5.50) for the devised reversible path. There are a large number of paths by which the gas can be taken from the initial state to the final state. Let us consider two such convenient paths 1-A-2 and 1-B-2 as shown in Fig. 5.14 and evaluate the entropy change.

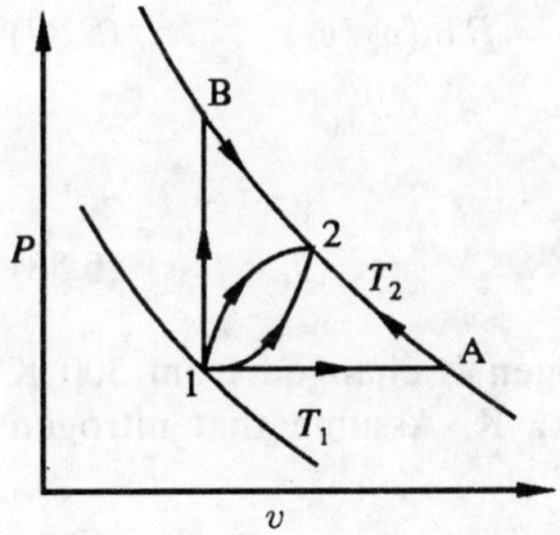

Fig. 5.14 *Two different reversible paths connecting the states 1 and 2 for an ideal gas.*

Path 1-A-2 Constant pressure heating from state 1 to state A, followed by reversible isothermal process A-2.

$$s_A - s_1 = \int_1^A \frac{dq}{T} = \int_1^A \frac{dh}{T} = \int_1^A C_p \frac{dT}{T} = C_p \ln \frac{T_2}{T_1} \tag{5.51}$$

$$s_2 - s_A = \int_A^2 \frac{dq}{T} = \int_A^2 \frac{du + Pdv}{T} = \int_A^2 \frac{Pdv}{T}$$

(since $du = 0$ for an ideal gas at constant temperature)

or

$$s_2 - s_A = \int_A^2 R \frac{dv}{v} = R \ln \frac{v_2}{v_A} \tag{5.52}$$

At constant temperature

$$P_2 v_2 = P_A v_A \quad \text{or} \quad \frac{v_2}{v_A} = \frac{P_A}{P_2} = \frac{P_1}{P_2} \tag{5.53}$$

Therefore,

$$s_2 - s_A = R\ln(P_1/P_2) = -R\ln(P_2/P_1) \tag{5.54}$$

Hence

$$s_2 - s_1 = (s_2 - s_A) + (s_A - s_1) = C_p \ln(T_2/T_1) - R\ln(P_2/P_1) \tag{5.55}$$

Path 1-B-2 Constant volume heating from T_1 to T_2 (1 to B) followed by reversible isothermal process B-2.

$$s_B - s_1 = \int_1^B \frac{dq}{T} = \int_1^B \frac{du}{T} = \int_1^B C_v \frac{dT}{T} = C_v \ln(T_2/T_1) \tag{5.56}$$

$$s_2 - s_B = \int_B^2 \frac{du + Pdv}{T} = \int_B^2 \frac{Pdv}{T} = \int_B^2 R\frac{dv}{v} = R\ln(v_2/v_B) = R\ln(v_2/v_1) \tag{5.57}$$

Therefore,

$$s_2 - s_1 = (s_2 - s_B) + (s_B - s_1) = C_v \ln(T_2/T_1) + R\ln(v_2/v_1) \tag{5.58}$$

EXAMPLE 5.11 Calculate the entropy change when 1 kg nitrogen is changed from 300 K and 1 bar to 500 K and 3 bar. For nitrogen, C_p = 1.0416 kJ/ kg K. Assume that nitrogen behaves like an ideal gas.

SOLUTION The entropy change of an ideal gas is given by

$$\Delta s = C_p \ln(T_2/T_1) - R\ln(P_2/P_1)$$

For nitrogen,

$$R = \frac{R_u}{M} = \frac{8.314 \times 10^3}{28}$$

$$= 0.2969 \text{ kJ/kg K}$$

Therefore,

$$\Delta s = 1.0416 \times 10^3 \ln(500/300) - 0.2969 \times 10^3 \ln(3/1)$$

$$= 0.2059 \text{ kJ/kg K}$$

5.11 Principle of Entropy Increase

Consider a system which is undergoing a reversible cycle 1-A-2-B-1 as shown in Fig. 5.15. The Clausius inequality for the reversible cycle gives

$$\int_{1A2B1} \frac{dq}{T} = \int_{1A2} \frac{dq}{T} + \int_{2B1} \frac{dq}{T} = 0 \tag{5.59}$$

Suppose the system undergoes an irreversible cycle 1-A-2-C-1 in which the path 1-A-2 is reversible and the path 2-C-1 is irreversible. The Clausius inequality for the irreversible cycle gives

$$\int \left(\frac{dq}{T}\right)_{1A2C1} = \int \left(\frac{dq}{T}\right)_{1A2} + \int \left(\frac{dq}{T}\right)_{2C1} < 0 \tag{5.60}$$

Subtracting Eq. (5.60) from Eq. (5.59), we get

$$\int_{2B1} \frac{dq}{T} - \int_{2C1} \frac{dq}{T} > 0$$

or $$\int \left(\frac{dq}{T}\right)_{2B1} = \int_2^1 ds > \int \left(\frac{dq}{T}\right)_{2C1}$$

since 2-B-1 is a reversible path, $ds = (dq/T)_R$

or, for an irreversible process

$$ds > \frac{dq}{T} \tag{5.61}$$

One can generalize and write

$$ds \geq \frac{dq}{T} \tag{5.62}$$

where the equality sign holds good for reversible processes and the inequality sign holds good for irreversible processes.

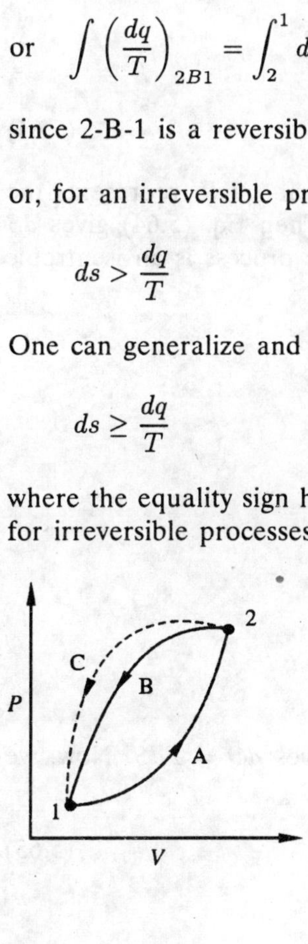

Fig. 5.15 A reversible cycle 1-A-2-B-1 and an irreversible cycle 1-A-2-C-1.

Now let us consider an isolated system which undergoes a change of state. Then for the isolated system $dq = 0$ and hence Eq. (5.62) gives

$$ds \geq 0 \text{ for an isolated system.} \tag{5.63}$$

Equation (5.63) tells that the entropy of an isolated system either increases or in the limit remains constant. If there is any irreversibility, the entropy of an isolated system increases. This fact is usually stated as the principle of entropy increase.

The principle of entropy increase dictates the direction of the process. No process can occur in the direction of entropy decrease for an isolated system. For an isolated system,

$\Delta s = 0$ For reversible process

$\Delta s > 0$ For irreversible process

5.12 Temperature–Entropy Diagram

The entropy change of a system is given by

$$dS = (dQ/T)_R$$

or

$$\int dQ = \int TdS \tag{5.64}$$

If T and S are chosen as the independent variables, then the integral $\int TdS$ represents the area under the curve. For a reversible adiabatic process $dQ = 0$. Then Eq. (5.64) gives $dS = 0$ for a reversible adiabatic process. That is, a reversible adiabatic process is an isentropic process. The Carnot cycle on a T–S diagram is shown in Fig. 5.16.

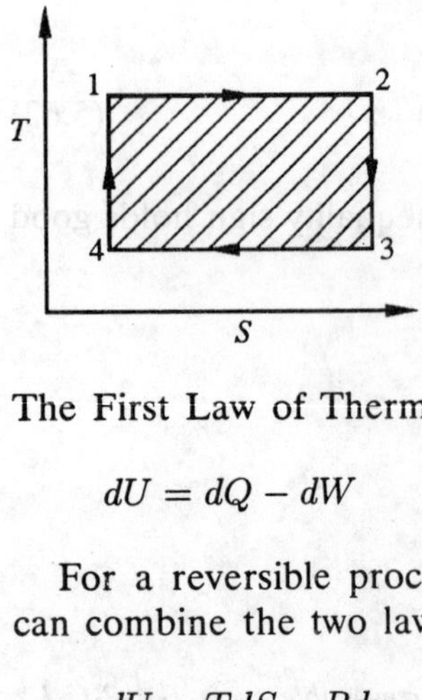

Fig. 5.16 Carnot cycle on T–S diagram.

The First Law of Thermodynamics is

$$dU = dQ - dW$$

For a reversible process the Second Law of Thermodynamics gives $dQ = TdS$. Now, we can combine the two laws of thermodynamics and write

$$dU = TdS - Pdv \tag{5.65}$$

Equation (5.65) is valid for all processes—reversible as well as irreversible—since it tells us that $U = U(S, v)$. Since U is a thermodynamic property, its change does not depend on the path, and the change in U is the same whether the system undergoes a reversible or irreversible process.

The First Law of Thermodynamics when applied to a cyclic change gives

$$\oint dU = 0 = \oint TdS - \oint Pdv$$

or

$$\oint TdS = \oint Pdv \tag{5.66}$$

Therefore, on a T–S diagram, the area enclosed by a cycle gives the net work done (shaded area in Fig. 5.16).

5.13 Second Law Analysis of a Control Volume

Consider the control volume shown in Fig. 5.17. Let

$m(t)$ = mass inside CV at time t

$s(t)$ = specific entropy of mass in CV at time t

$\dot{m}_i, \dot{m}_e$ = mass flow rate at inlet and exit, respectively

s_i, s_e = specific entropy of mass at inlet and exit, respectively

Q = rate of heat flow into CV

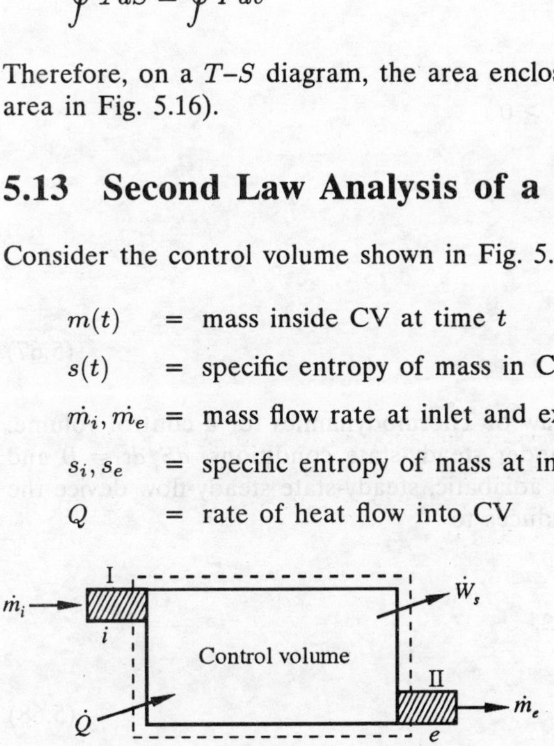

Fig. 5.17 Control volume for second law analysis.

We shall adopt the procedure used in applying the First Law of Thermodynamics to a control volume. At time t, the system is the combination of CV and the region I and at time $t + dt$, the system is the combination of CV and the region II.

Entropy of the system at time t

$$= m(t)s(t) + \dot{m}_i s_i dt = S(t) + \dot{m}_i s_i dt$$

Entropy of the system at time $(t + dt)$

$$= m(t + dt)s(t + dt) + \dot{m}_e s_e dt = S(t + dt) + \dot{m}_e s_e dt$$

Therefore, entropy change of the system is given by

$$(\Delta S)_{sys} = [S(t+dt) + \dot{m}_e s_e dt] - [S(t) + \dot{m}_i s_i dt]$$

Entropy change of the surroundings is given by

$$(\Delta S)_{Sur} = \frac{-\dot{Q}dt}{T}$$

The combination of system and surroundings is an isolated system. The principle of entropy increase gives $\Delta S \geq 0$ for an isolated system.

Therefore, $(\Delta S)_{Sys} + (\Delta S)_{Sur} \geq 0$

or $$[S(t+dt) + \dot{m}_e s_e dt] - [S(t) + \dot{m}_i s_i dt] - \frac{\dot{Q}dt}{T} \geq 0$$

or $$\frac{S(t+dt) - S(t)}{dt} + \dot{m}_e s_e - \dot{m}_i s_i - \frac{\dot{Q}}{T} \geq 0$$

or

$$\frac{dS}{dt} + \dot{m}_e s_e - \dot{m}_i s_i - \frac{\dot{Q}}{T} \geq 0 \tag{5.67}$$

which is the general expression of the Second Law of Thermodynamics for a control volume.

If the device (control volume) is operating under steady-state conditions, $dS/dt = 0$ and for an adiabatic device $\dot{Q} = 0$. Therefore, for an adiabatic, steady-state steady-flow device the Second Law of Thermodynamics (Eq. (5.67)) reduces to

$$\dot{m}_e s_e - \dot{m}_i s_i \geq 0 \quad \text{or} \quad \dot{m}(s_e - s_i) \geq 0.$$

or

$$s_e \geq s_i \tag{5.68}$$

That is, the exit entropy is greater than or equal to the entropy at the inlet. In Eq. (5.68) the equality sign holds good for reversible processes and the inequality sign holds good for irreversible processes. For a reversible, adiabatic device operating under steady-state conditions

$$s_e = s_i \tag{5.69}$$

EXAMPLE 5.12 An inventor claims to have designed a device which takes in 2 kmol/s of air (ideal gas, $\gamma = 1.4$) at 2.5 bar and 350 K and delivers equal amounts of hot air and cold air at 1 bar. The hot air leaves the device at 450 K while the cold air leaves the device at 250 K. He further claims that the device does not require any energy input in the form of work or heat. Judge whether the claim is justified or not.

SOLUTION A schematic representation of the device is shown in Fig. 5.18.

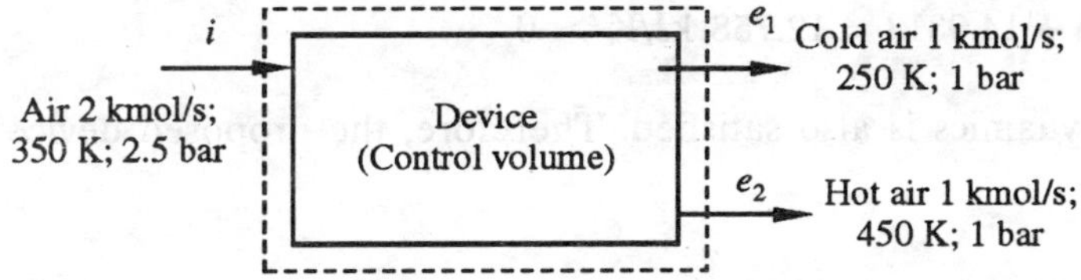

Fig. 5.18 Sketch for Example 5.12.

The device is feasible if it does not violate the laws of thermodynamics. Therefore, let us verify whether the first and second laws of thermodynamics are satisfied or not. The First Law of Thermodynamics for the control volume (adiabatic, steady-state, steady-flow) gives

$$\dot{m}_i h_i = \dot{m}_{e1} h_{e1} + \dot{m}_{e2} h_{e2}$$

$$\dot{m}_i = 2 \text{ kmol/s}; \quad \dot{m}_{e1} = \dot{m}_{e2} = 1 \text{ kmol/s}$$

$$2h_i = h_{e1} + h_{e2}$$

$$2C_p T_i = C_p(T_{e1} + T_{e2})$$

$$2 \times 350 = 250 + 450 \tag{A}$$

Equation (A) is satisfied. That is, the First Law of Thermodynamics is satisfied. The Second Law of Thermodynamics gives

$$\dot{m}_{e1} s_{e1} + \dot{m}_{e2} s_{e2} \ge \dot{m}_i s_i \quad \text{or} \quad s_{e1} + s_{e2} \ge 2s_i$$

or

$$(s_{e1} - s_i) + (s_{e2} - s_i) \ge 0 \tag{B}$$

For an ideal gas, the change in entropy is given by

$$\Delta s = C_p \ln(T_2/T_1) - R \ln(P_2/P_1)$$

We know that for an ideal gas

$$C_p = \frac{R\gamma}{\gamma - 1} \quad \text{or} \quad C_p = \frac{8.314 \times 10^3 \times 1.4}{1.4 - 1} = 29.1 \text{ kJ/kmol K}$$

Therefore,

$$s_{e1} - s_i = 29.1 \ln(250/350) - 8.314 \ln(1/2.5) = -2.1733 \text{ kJ/kmol K}$$

$$s_{e2} - s_i = 29.1 \ln(450/350) - 8.314 \ln(1/2.5) = -14.9313 \text{ kJ/kmol K}$$

$$(s_{e1} - s_i) + (s_{e2} - s_i) = -2.1733 + 14.9313 = 12.758 \text{ kJ/K} > 0$$

Hence, the Second Law of Thermodynamics is also satisfied. Therefore, the proposed device is theoretically feasible.

5.14 Available and Unavailable Energy

We have learnt that the First Law of Thermodynamics relates the change in the energy of the system to the heat and work interactions across the boundaries of the system. However, the First Law of Thermodynamics does not specify the amount of work that can be done by a system when it receives a certain amount of energy as heat. The Second Law of Thermodynamics tells us that it is not possible to convert all the heat absorbed by a system into work. It tells us that in a heat engine (cyclically operating device), part of the energy received as heat has to be discarded to a sink at low temperature. Suppose a certain quantity of energy Q as heat can be received from a body at temperature T. How much of this energy can be converted into work or what is the energy available for conversion into work? The maximum work can be obtained by operating a Carnot engine (reversible engine) using the body at T as the source and the ambient atmosphere at T_0 as the sink. Then the work that can be obtained is given by

$$W = Q\eta = Q\left(1 - \frac{T_0}{T}\right) = Q - T_0|\Delta s| \tag{5.70}$$

where ΔS is the change in the entropy of the body supplying the energy as heat. Therefore, the available energy is given by $Q - T_0|\Delta S|$. The Carnot cycle and the available energy is shown in Fig. 5.19. The area 1-2-3-4 in Fig. 5.19 represents the available energy. The shaded area 4-3-B-A represents the energy which is discarded to the ambient atmosphere and this quantity of energy cannot be converted into work and is called *unavailable energy*.

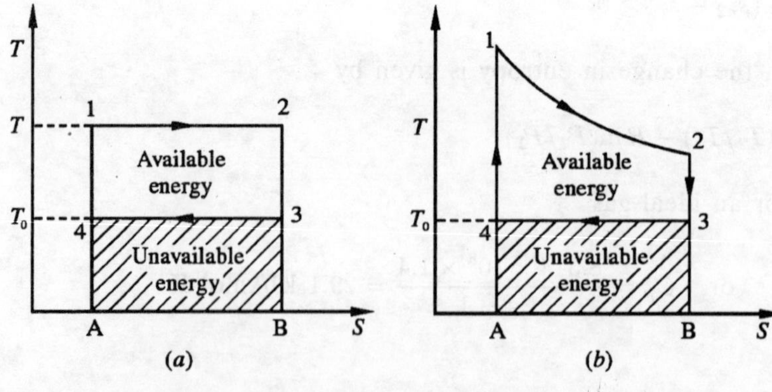

Fig. 5.19 *Available and unavailable energy (a) when a constant temperature thermal reservoir is used as a source. (b) when a finite body is used as a source.*

Suppose a finite body is used as a source. As energy is withdrawn from the body, its temperature keeps on decreasing. For this purpose we can use a large number of differential Carnot engines with the given body as the source. The differential work done by the engine is given by

$$dW = dQ\eta = dQ\left(1 - \frac{T_0}{T}\right)$$

If the initial and final temperatures of the source are T_1 and T_2 respectively, the total work done or the available energy is given by

$$W = \int dQ\eta = \int_{T_1}^{T_2} dQ\left(1 - \frac{T_0}{T}\right) = Q - T_0 \int_{T_1}^{T_2} \frac{dQ}{T}$$

or $\quad W = Q - T_0|\Delta S|$

5.15 Loss in Available Energy

Suppose a certain quantity of energy Q is transferred from a body at constant temperature T_1 to another body at constant temperature T_2 ($T_2 < T_1$). This transfer of energy takes place with finite temperature difference between the two interacting bodies and is irreversible. This irreversibility leads to a loss in available energy.

Initial available energy, with the body at T_1

$$= Q\left(1 - \frac{T_0}{T_1}\right)$$

Final available energy, with the body at T_2

$$= Q\left(1 - \frac{T_0}{T_2}\right)$$

Loss in available energy

$$= Q\left(1 - \frac{T_0}{T_1}\right) - Q\left(1 - \frac{T_0}{T_2}\right) = T_0\left(\frac{Q}{T_2} - \frac{Q}{T_1}\right) = T_0 \Delta S_{\text{uni}} \tag{5.71}$$

where ΔS_{uni} is the change in the entropy of the universe. Equation (5.71) shows that whenever $\Delta S_{\text{uni}} > 0$, there is a loss in available energy.

EXAMPLE 5.13 Saturated steam at 100°C is available in a particular plant and it is required to use saturated liquid water at 100°C in another unit of the same plant. Calculate the available energy per kg of steam if steam is condensed to liquid water at the same temperature. The latent heat of vaporization of water at 100°C is 2257 kJ/kg. The temperature of the surroundings is 25°C.

SOLUTION The initial state is saturated steam at 100°C and the final state is saturated liquid at 100°C. The phase transition from vapor to liquid takes place at constant temperature and pressure. Therefore,

$$q = \Delta h = 2257 \text{ kJ/kg}$$

$$\text{Available energy} = q - T_0|\Delta s| = q - T_0\frac{\Delta h}{T} = 2257 \times 10^3 - 298 \times \frac{2257 \times 10^3}{373} = 453.82 \text{ kJ/kg}$$

EXAMPLE 5.14 Two identical blocks of metal, each having a mass of 100 kg and a specific heat of 0.5 kJ/kg K are available. One block is at 1200 K while the other block is at 600 K. The two blocks are brought into thermal contact with each other and allowed to reach a state of thermal equilibrium. Calculate the loss in available energy associated with this process, if the ambient temperature is 300 K.

SOLUTION Let T_1, T_2 = initial temperature of blocks 1 and 2, respectively. Final equilibrium temperature T_f is given by

$$= \frac{1200 + 600}{2} = 900 \text{ K}$$

$$\text{Loss in available energy} = T_0\Delta S_{\text{uni}} = T_0\left[mC_p \ln\frac{T_f}{T_1} + mC_p \ln\frac{T_f}{T_2}\right]$$

$$= 300\left[100 \times 0.5 \ln\left(\frac{900}{1200}\right) + 100 \times 0.5 \ln\left(\frac{900}{600}\right)\right] = 1766.75 \text{ kJ}$$

Review Questions

1. What are the limitations of the First Law of Thermodynamics?
2. What is a heat engine? Give two examples of heat engines.
3. What is a thermal reservoir?
4. What is a refrigerator and what is a heat pump? How do they differ?
5. Define the thermal efficiency of a heat engine.
6. Define the coefficient of performance (COP) of a refrigerator and the COP of a heat pump.
7. How is the COP of a heat pump related to the COP of a refrigerator when both the units are operating between the same set of thermal reservoirs?
8. State the Kelvin–Planck and Clausius statements of the Second Law of Thermodynamics.
9. What is a Perpetual Motion Machine of the Second Kind?
10. Does the Clausius statement prohibit the transfer of energy as heat from a low temperature body to a high temperature body? Does the domestic refrigerator violate the Second Law of Thermodynamics? Explain.
11. Prove that violation of the Kelvin–Planck statement leads to violation of the Clausius statement.
12. Show that violation of the Clausius statement leads to violation of the Kelvin–Planck statement.

13. How can one prove the equivalence of the Kelvin–Planck and Clausius statements of the Second Law of Thermodynamics?
14. Can a spontaneous process proceed in both directions?
15. What is the essence of the Second Law of Thermodynamics and why is it called a directional law?
16. What is the objective of a refrigerator and what is the objective of a heat pump?
17. Can the thermal efficiency of a heat engine be equal to one?
18. Can the COP of a refrigerator or a heat pump be equal to infinity?
19. Sketch a Carnot cycle on the P–v diagram and name the processes involved in the cycle.
20. Derive a relation to estimate the efficiency of a Carnot engine which uses an ideal gas as the working fluid.
21. State the Carnot theorems and prove them.
22. Is it possible to increase the efficiency of a reversible engine by changing the working fluid? Explain.
23. Name the factors on which the efficiency of a reversible heat engine depends.
24. What is the thermodynamic temperature scale and how is it established?
25. How many fixed points are used in establishing the thermodynamic temperature scale and what are they?
26. What is the Clausius inequality?
27. In a given process is it possible to convert all the energy absorbed as heat from a single thermal reservoir into work? If possible, give an example.
28. Explain the method of calculating the entropy change if a system undergoes a change of state by an irreversible process.
29. What is the entropy change if an ideal gas at T_1 and P_1 is changed to T_2 and P_2?
30. What is the entropy change if an ideal gas undergoes a state change from T_1 and v_1 to T_2 and v_2?
31. Develop a relation to estimate the entropy change for an ideal gas.
32. State the principle of entropy increase.
33. Can the entropy of a system decrease? Explain.
34. How can you judge the feasibility of a process based on entropy change?
35. Show that a reversible-adiabatic process is an isentropic process.
36. State the combined statement of the two laws of thermodynamics. Is it applicable to all processes? Explain.
37. Develop the general second law expression for a control volume.
38. What is available energy?
39. If energy transfer as heat occurs with a finite temperature difference, does the available energy remain constant? Explain.
40. How is the loss in available energy related to the change in the entropy of the universe?

Problems

5.1 A system receives 100 kJ energy as heat from a single thermal reservoir and does 70 kJ work. Is it possible to restore the system to its initial state by following an adiabatic path?

5.2 Prove that two reversible adiabatic lines cannot intersect.

5.3 A Carnot engine receives 100 kJ energy as heat from a reservoir at 1000 K and rejects energy as heat to a sink at 300 K. Calculate the work done by the engine and the quantity of energy rejected to the sink.

5.4 It is desired to operate a Carnot engine which uses a source at 727°C and a sink at 27°C. If the desired power output of the engine is 50 kW, how much energy does it absorb per second from the source?

5.5 A Carnot engine A absorbs 1000 kJ of energy as heat from a thermal reservoir at 1200 K and rejects energy as heat to a thermal reservoir at T. Another Carnot engine B absorbs the same quantity of energy as rejected by engine A, from the thermal reservoir at T and rejects energy as heat to a thermal reservoir at 300 K. If the efficiencies of the engines A and B are equal, calculate the work done by engine A and by engine B.

5.6 Estimate the coefficient of performance of a Carnot refrigerator operating between 40°C and –20°C.

5.7 Calculate the COP of a heat pump operating between 25°C and 5°C.

5.8 The thermal efficiency of a Carnot engine can be increased either (a) by increasing the source temperature while the sink is maintained at constant temperature or (b) by decreasing the sink temperature while the source is held at constant temperature. Which of the above two possibilities is more effective for a given change in the temperature of the reservoir?

5.9 Two reversible engines are operating between three thermal reservoirs at temperatures T_1, T_2 and T_3. The first engine operates between the reservoirs at T_1 and T_2 while the second engine uses the reservoirs at T_2 and T_3 ($T_1 > T_2 > T_3$). The second engine absorbs the amount of energy from the reservoir at T_2 which is equal to the quantity of energy rejected by the first engine to the reservoir at T_2, so that there is no change in the energy of the reservoir at T_2. Determine the relation between T_1, T_2 and T_3 if (a) the thermal efficiencies of both the engines are the same (b) the work delivered by both the engines is the same.

5.10 The coefficient of performance of a refrigerator can be increased either (a) by increasing the temperature of the low temperature reservoir while the temperature of the high temperature reservoir is held constant or (b) by decreasing the temperature of the high temperature reservoir while the temperature of the low temperature reservoir is held constant. Which of the above two possibilities is more effective for a given change in the temperature of the reservoir?

5.11 An inventor claims to have developed an engine with a thermal efficiency of 0.8. He further claims that the engine will receive energy as heat from a source at 800°C and use the atmosphere at 27°C as the sink. Would you support this claim?

5.12 A Carnot engine with a thermal efficiency of 0.6 is used to operate a Carnot refrigerator with a COP of 5. Both the engine and the refrigerator reject energy to the surrounding atmosphere. Calculate the energy absorbed by the engine as heat from the source for every kJ of energy absorbed as heat from the cold space by the refrigerator.

5.13 A Carnot refrigerator is to be used to maintain certain drugs at –20°C while the surrounding atmosphere is at 37°C. The heat leak rate into the cold space of the refrigerator from the surroundings is estimated at 0.2 kW. Estimate the power required to operate the refrigerator.

5.14 It is required to maintain an auditorium at 25°C in summer when the surrounding atmosphere is at 43°C. If the heat generation in the auditorium is 70 kW, because of the persons occupying it, estimate the power needed to operate the refrigerating plant.

5.15 It is desired to maintain an auditorium at 30°C in winter while the surrounding atmosphere is at 5°C. The rate of heat losses from the walls and roof is estimated at 60 kW. Calculate the power required to operate a heat pump.

5.16 A reversible heat engine operating with a source at 427°C and the surrounding atmosphere at 27°C as the sink is used to drive a refrigerator which maintains a cold space at −27°C and rejects energy as heat to the surrounding atmosphere. Estimate the ratio of the energy absorbed by the heat engine to the energy absorbed by the refrigerator from the cold space.

5.17 An inventor claims to have designed a refrigerator which extracts energy as heat at a rate of 0.4 kW from a cold space at −20°C, and consumes energy at a rate of 0.06 kW only. Assuming that the surrounding atmosphere is at 27°C, judge whether the claim is justified or not.

5.18 A device has been designed to be used as a heat pump in winter and as a refrigerating unit in summer to maintain a house at 25°C throughout the year. The energy losses as heat from the walls and roof are estimated at 2 kW per degree Celsius temperature difference between the surrounding atmosphere and the inside space of the house. If the surrounding atmosphere temperature is 37°C in summer and 7°C in winter, estimate the power required to operate the unit in summer and in winter.

5.19 It was claimed that a heat engine operates between three thermal reservoirs. It absorbs 600 kJ energy as heat from a reservoir at 600 K and rejects 50 kJ and 100 kJ energy as heat to reservoirs at 400 K and 300 K, respectively, and delivers 450 kJ work. Determine whether the heat engine is feasible or not.

5.20 The latent heat of vaporization of water is 2114.3 kJ/kg and 1940.7 kJ/kg at 150°C and 200°C, respectively. Calculate the entropy change due to vaporization at 150°C and 200°C.

5.21 The entropy of saturated liquid water at 100°C (and one bar pressure) is 1.3069 kJ/kg K. Assuming that the specific heat at constant pressure for steam is 2 kJ/kg K and the latent heat of vaporization of water at 100°C is 2257 kJ/kg, calculate the entropy of steam at 3 bar and 250°C. Assume the steam to behave like an ideal gas.

5.22 Show that

$$\frac{T_2}{T_1} = \left(\frac{P_2}{P_1}\right)^{(\gamma-1)/\gamma}$$

for an ideal gas undergoing an isentropic process.

5.23 Calculate the entropy change if a mass m of liquid at temperature T_1 is mixed with an equal amount of the same liquid at temperature T_2. The specific heat of the liquid is constant and is equal to C_p.

5.24 Two identical bodies of mass m and specific heat C_p are initially at temperatures T_1 and T_2 ($T_1 > T_2$). Estimate the maximum work that can be obtained, if the bodies are used as a source and a sink to operate a heat engine.

5.25 Air (ideal gas, $\gamma = 1.4$), is steadily flowing through a pipeline. The temperature and pressure are measured at two different stations A and B and it is found that $T_A = 50°C$, $T_B = 30°C$; $P_A = 150$ kPa and $P_B = 120$ kPa. Determine the direction of flow of air.

6 Properties of Steam

We have studied the first and second laws of thermodynamics and are in a position to carry out a thermodynamic analysis of various processes. To evaluate the heat and work interactions across the boundary of a system, it is essential to have a knowledge of the thermodynamic properties of the system under consideration. Since steam is extensively used in many industries, we shall study the properties of steam in this chapter.

6.1 States of Water

Substances can be found in different physical states like solid, liquid and gas. The different physical states of H_2O are ice, water and steam. All substances—solids, liquids and gases—change their specific volume when they are subjected to a change in pressure and temperature. Depending on the nature of the substance and the applied pressure and temperature, the change in the specific volume may be large or negligible. If the change in the specific volume is negligible, as in the case of solids and liquids, the substance is treated as incompressible. The isothermal compressibility of a substance (κ) is defined as

$$\kappa = -\frac{1}{v}\left(\frac{\partial v}{\partial P}\right)_T \tag{6.1}$$

The specific volume of a gas changes to a large extent with temperature and pressure. Hence, gases are treated as compressible substances.

Consider a certain quantity of liquid water at 25°C enclosed in a frictionless piston-cylinder assembly as shown in Fig. 6.1(a). The piston is massless and is exposed to atmospheric pressure. Suppose energy is transferred as heat to the water while the pressure is held constant, the temperature and specific volume of the water increase and follow the path 1-2 as shown in Fig. 6.1(b).

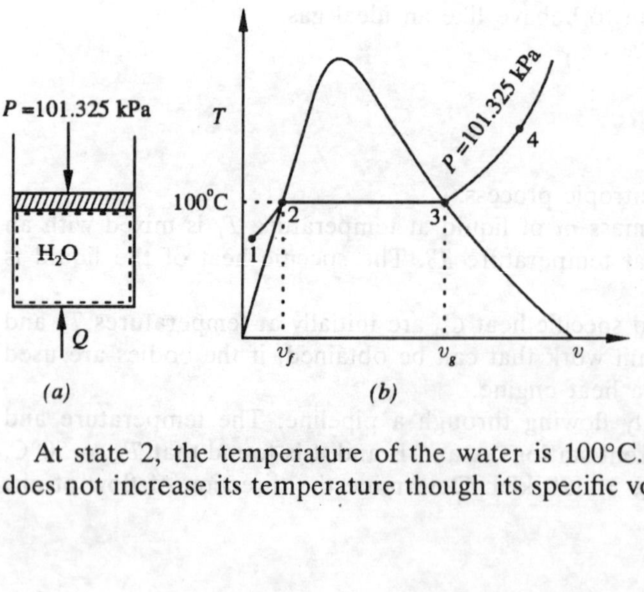

Fig. 6.1 *(a) Water (System) in a piston-cylinder assembly held at constant pressure (b) T–v diagram showing the path followed by water during constant pressure heating.*

At state 2, the temperature of the water is 100°C. Further addition of heat to the system does not increase its temperature though its specific volume increases. At state 2, vaporization starts and some of the liquid converts into vapor. The temperature of the vapor will be the same as that of the liquid and these two phases will be in a state of equilibrium. Further addition of heat to the system moves it along the path 2-3. At state 2, the liquid is at its boiling point and at state 3 all the liquid is converted into vapor. At state 3, the vapor is at its condensing point. A state where a phase change begins or ends is called a *saturated state*. Therefore, state 2 represents saturated liquid and its specific volume is v_f, and state 3 represents saturated vapor and its specific volume is v_g.

At a pressure of 101.325 kPa, the phase change for water takes place at 100°C. The temperature at the saturated state is called *saturation temperature*. That is, for water at 101.325 kPa, the saturation temperature is 100°C. At a specified temperature, the pressure at which the phase change takes place is called *saturation pressure*. Therefore, the saturation pressure of water at 100°C is 101.325 kPa.

On continued heating, the system moves along the path 3-4 and at state 4, the temperature of the vapor is greater than the saturation temperature. The steam in state 4 is said to be in the superheated state. The liquid water at state 1 is at a temperature which is less than the saturation temperature at the given pressure. The liquid, at a temperature less than the saturation temperature at a given pressure is said to be a *subcooled liquid*. A liquid at a pressure which is greater than the saturation pressure at a given temperature is said to be a *compressed liquid*. The saturation pressure of water at 25°C is 0.0317 bar. Therefore, the liquid at 25°C and 101.325 kPa can be called a compressed liquid at 25°C.

6.2 Temperature–Volume Diagram

If the heating experiment described above is performed at different pressures, one obtains the temperature–volume diagram which appears as in Fig. 6.2. The locus of all the saturated liquid states gives the saturated liquid curve AC and the locus of all the saturated vapor states gives the saturated vapor curve BC.

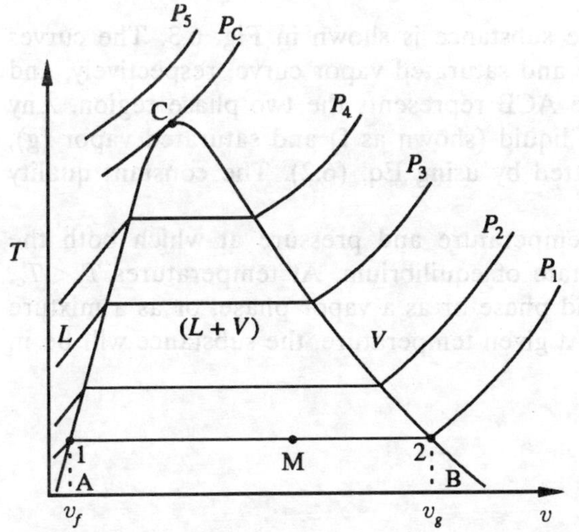

Fig. 6.2 *T–v diagram for a pure substance.* $P_5 > P_C > P_4 > P_3 > P_2 > P_1$.

The point C represents the critical point. It can be observed from Fig. 6.2 that the difference between v_g and v_f reduces as the pressure is increased, and at the critical point, $v_f = v_g$. At the critical point, the pressure is called the *critical pressure* P_c, and the temperature is called the *critical temperature* T_c. At the critical point, the two phases—liquid and vapor—are indistinguishable. The region under the curve ACB is the two-phase region, where both liquid and vapor are present. The region to the left of the curve AC represents the liquid region and the region to the right of the curve BC is the vapor region.

The point M in the two-phase region represents a mixture of saturated liquid and saturated vapor. Let

m_f = mass of saturated liquid $\qquad$ m_g = mass of saturated vapor

$m = m_f + m_g$ = mass of mixture $\qquad$ v = specific volume of mixture

Then, the total volume of the mixture is given by

$$V = m_f v_f + m_g v_g \quad \text{or} \quad v = \frac{V}{m} = \frac{m_f}{m} v_f + \frac{m_g}{m} v_g$$

or

$$v = (1 - X) v_f + X v_g \tag{6.2}$$

where

$$X = \frac{m_g}{m} = \frac{\text{mass of vapor}}{\text{mass of mixture}} = \text{quality} \tag{6.3}$$

6.3 Pressure–Volume Diagram

The pressure–volume (P–V) diagram for a pure substance is shown in Fig. 6.3. The curves AC and BC represent the saturated liquid curve and saturated vapor curve, respectively, and C is the critical point. The area under the curve ACB represents the two-phase region. Any point M in this region is a mixture of saturated liquid (shown as f) and saturated vapor (g), and the quality X of the mixture can be estimated by using Eq. (6.2). The constant quality lines and isotherms are also shown in Fig. 6.3.

The *critical point* C represents the highest temperature and pressure at which both the liquid phase and vapor phase can coexist in a state of equilibrium. At temperatures $T < T_c$, (say T_1), the substance can exist either as a liquid phase or as a vapor phase, or as a mixture of liquid and vapor, depending on the pressure. At given temperature, the substance will be in

Liquid phase if $P > P^s$
Vapor phase if $P < P^s$
Liquid and vapor if $P = P^s$

where P^s is the saturation pressure at the given temperature.

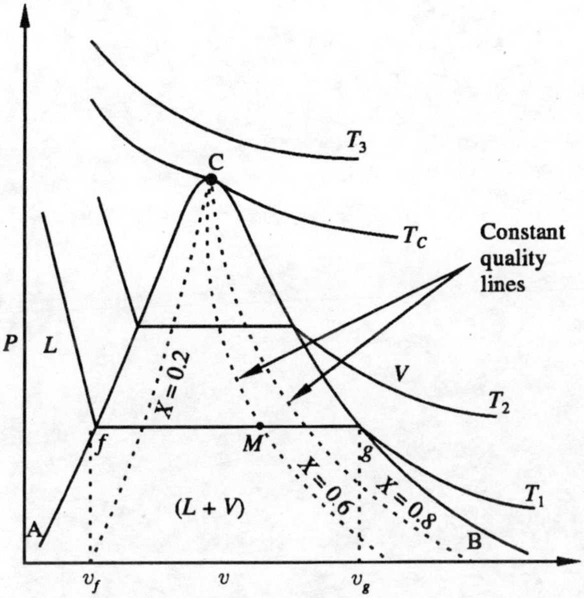

Fig. 6.3 *P–v diagram for a pure substance.* $T_3 > T_c > T_2 > T_1$.

6.4 Temperature–Entropy Diagram

The temperature–entropy (*T–s* diagram for a pure substance is shown in Fig. 6.4 in which the constant quality lines and constant pressure lines are indicated.

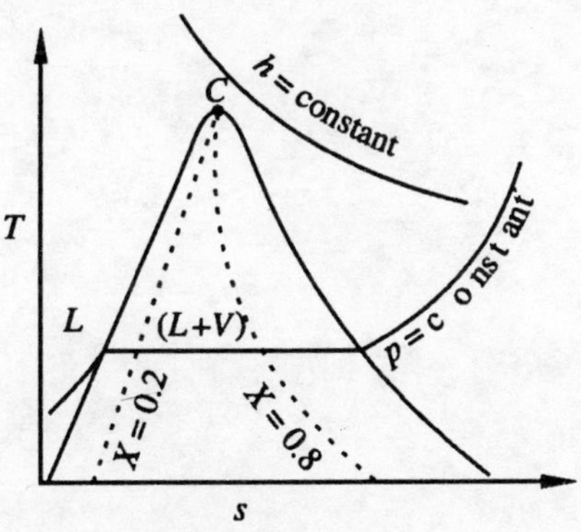

Fig. 6.4 *T–s diagram for a pure substance.*

6.5 Mollier (*h–s*) Diagram

The thermodynamic properties of a substance are quite often shown on a Mollier diagram. The *h–s* diagram was introduced by Richard Mollier (1863–1935) and was named Mollier diagram after him. A typical Mollier diagram showing the essential features is presented in Fig. 6.5. It consists of a family of constant pressure lines, constant temperature lines and constant volume lines plotted on enthalpy versus entropy coordinates. In the two-phase region, the constant pressure and constant temperature lines coincide. In thermodynamics, we deal

Fig. 6.5 Mollier diagram.

with isobaric processes, isothermal processes and isentropic processes. All these processes can be conveniently analyzed with the help of a Mollier diagram.

6.6 Steam Tables

When the thermodynamic data of substances is presented in the form of diagrams, the values cannot be read accurately and at times it is necessary to interpolate the data. If the same data is presented in the form of tables, some of these drawbacks can be eliminated. When the data is presented in the form of a table, the data can be read accurately and there is no limitation on the number of properties which can be reported. The important thermodynamic properties like T, P, v, u, h and s for saturated liquid, saturated vapor and superheated vapor are usually provided in the form of tables. Such tables are available for several substances. In general, all these tables are similar. Therefore, it is enough to present here representative tables for one substance, that is, for water.

Since the properties like internal energy, enthalpy and entropy of a system cannot be directly measured, they are related to the change in the energy of the system. Hence, one can determine $\Delta u, \Delta h$ and Δs, but not the absolute values of these properties. Therefore, it is necessary to choose a reference state to which these properties are arbitrarily assigned some numerical values. For water, the triple point ($T = 0.01°C$ and $P = 0.6113$ kPa) is selected as the reference state, where the internal energy and entropy of the saturated liquid are assigned a zero value. This choice of reference state is immaterial because we are only interested in the change in the properties of a substance while it undergoes a process.

The properties of steam are given in three different tables:

- Saturated steam (temperature) table
- Saturated steam (pressure) table
- Superheated steam.

In the saturated steam tables, the properties of saturated liquid which is in equilibrium with saturated vapor, are presented. During phase transition, the pressure and temperature are not independent of each other. If the temperature is specified, the pressure at which both the phases coexist in equilibrium is equal to the saturation pressure. Similarly, if the pressure is specified, the saturation temperature has a definite value. Hence, it is possible to choose either temperature or pressure as the independent variable, to specify the state of a two-phase system. If the temperature is chosen as the independent variable, and the thermodynamic properties are tabulated, then the table is called the saturated steam (temperature) table. On the other hand, if the pressure is chosen as the independent variable, and the properties are tabulated, then the table is called the saturated steam (pressure) table. The two phases—liquid and vapor—can coexist in a state of equilibrium only upto the critical point. Therefore, the listing of the thermodynamic properties of steam in the saturated steam tables ends at the critical point (374.15°C and 221.2 bar).

If the steam exists in only one phase (superheated steam), it is necessary to specify two independent variables, pressure and temperature, for the complete specification of the state. In the superheated steam tables, the properties—v, u, h and s—are tabulated from the saturation temperature to some temperature, for a given pressure. The steam tables are presented in the Appendix and are self-explanatory.

The thermodynamic properties of a liquid and vapor mixture can be evaluated in terms of its quality. In particular, the specific volume, specific internal energy, specific enthalpy and specific entropy of a mixture of quality X are given by

$$v = (1 - X)v_f + Xv_g \tag{6.2}$$

$$u = (1 - X)u_f + Xu_g \tag{6.4}$$

$$h = (1 - X)h_f + Xh_g = h_f + Xh_{fg} \tag{6.5}$$

$$s = (1 - X)s_f + Xs_g \tag{6.6}$$

where

$$h_{fg} = h_g - h_f = \text{latent heat of vaporization} \tag{6.7}$$

EXAMPLE 6.1 A piston-cylinder assembly contains 0.5 m³ of dry saturated steam at a pressure of 2 bar. Determine the temperature and mass of steam.

SOLUTION Since the steam is in the saturated state, the saturation temperature at P = 2 bar is 120.23°C (Read the value from the saturated steam (pressure) table. At this pressure, v_g = 0.8854 m³/kg. Therefore,

$$\text{Mass of steam, } m = \frac{v}{v_g} = \frac{0.5}{0.8854} = 0.565 \text{ kg}$$

EXAMPLE 6.2 A vessel of volume 1 m³ contains saturated steam of quality 0.85 at 2 bar. Calculate the temperature of the steam, the mass of liquid and the mass of vapor.

SOLUTION The steam is saturated at P = 2 bar. Read the following values from the saturated steam (pressure) table,

$$T = 120.23°\text{C}; \quad v_f = 0.001\ 060\ 8 \text{ m}^3/\text{kg} \quad \text{and} \quad v_g = 0.8854 \text{ m}^3/\text{kg}$$

Temperature of steam = 120.23°C.

Specific volume of steam is given by

$$v = Xv_g + (1 - X)v_f = 0.85 \times 0.8854 + 0.15 \times 0.001\ 060\ 8 = 0.7527 \text{ m}^3/\text{kg}$$

$$\text{Total mass of steam, } m = \frac{V}{v} = \frac{1}{0.7527} = 1.329 \text{ kg}$$

Mass of vapor, $m_g = mX = 1.329 \times 0.85 = 1.13$ kg

Mass of liquid, $m_f = m(1 - X) = 1.329 \times 0.15 = 0.199$ kg

EXAMPLE 6.3 A piston-cylinder assembly contains 1 m^3 of saturated steam of quality 0.85 at a pressure of 3 bar. This assembly is brought into contact with a thermal reservoir and 1500 kJ of energy is added as heat at constant pressure. Determine the final state of the steam and the work done by it.

SOLUTION Read the following values at P = 3 bar from the saturated steam (pressure) table.

$$v_f = 0.001\ 073\ 5\ \text{m}^3/\text{kg}; \quad v_g = 0.6056\ \text{m}^3/\text{kg}$$

$$h_f = 561.43\ \text{kJ/kg}; \quad h_g = 2724.7\ \text{kJ/kg}$$

Initial specific volume v_1 is given by

$$v_1 = Xv_g + (1 - X)v_f = 0.85 \times 0.6056 + 0.15 \times 0.001\ 073\ 5 = 0.5149\ \text{m}^3/\text{kg}$$

Initial specific enthalpy h_1 is given by

$$h_1 = (1 - X)h_f + Xh_g = 0.15 \times 561.43 + 0.85 \times 2724.7 = 2400.2\ \text{kJ/kg}$$

$$\text{Mass of steam, } m = \frac{V}{v} = \frac{1}{0.5149} = 1.942\ \text{kg}$$

For a constant pressure process

$$dQ = dH \quad \text{or} \quad Q = H_2 - H_1$$

$$H_2 - H_1 = m(h_2 - h_1) = 1500\ \text{kJ} \quad \text{or} \quad 1.942(h_2 - 2400.2) = 1500$$

or $\quad h_2 = 3172.6$ kJ/kg

Therefore, the final state of the steam corresponds to P_2 = 3 bar and h_2 = 3172.6 kJ/kg. Since $h_2 > h_g$, the steam is available as superheated steam. From the superheated steam tables, we find that at P = 3 bar, h = 3069.7 kJ/kg at 300°C and h = 3275.2 kJ/kg at 400°C. Therefore, we make a linear interpolation to find the temperature of steam.

$$3172.6 = 3069.7 + \left(\frac{3275.2 - 3069.7}{400 - 300}\right)\Delta T \quad \text{or} \quad \Delta T = 50°\text{C}.$$

Hence, final temperature = 300 + 50 = 350°C.

Specific volume at P = 3 bar and T = 350°C is v = 0.9534 m^3/kg (obtained by interpolation).

$$\text{Work done by steam} = mP(v_2 - v_1) = 1.942 \times 3 \times 10^5(0.9534 - 0.5149) = 255.47\ \text{kJ}$$

The P–v diagram for the process followed by steam is shown in Fig. 6.6.

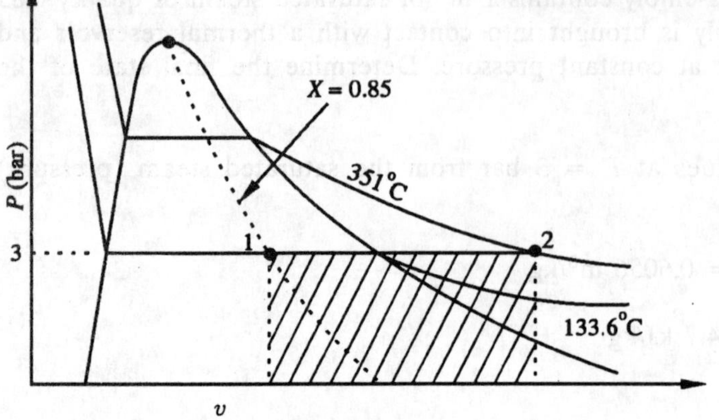

Fig. 6.6 Sketch for Example 6.3.

EXAMPLE 6.4 Rework Example 6.3 using the Mollier diagram.

SOLUTION The initial state of saturated steam is P = 3 bar and X = 0.85. Locate the initial state 1 at the intersection of P = 3 bar and X = 0.85 curves as shown in Fig. 6.7. Then read h_1 = 2400 kJ/kg and v_1 = 0.51 m^3/kg (obtained by interpolation between v = 0.5 m^3/kg and v = 0.75 m^3/kg curves). Then mass of steam,

$$m = V/v = 1/0.51 = 1.961 \text{ kg}$$

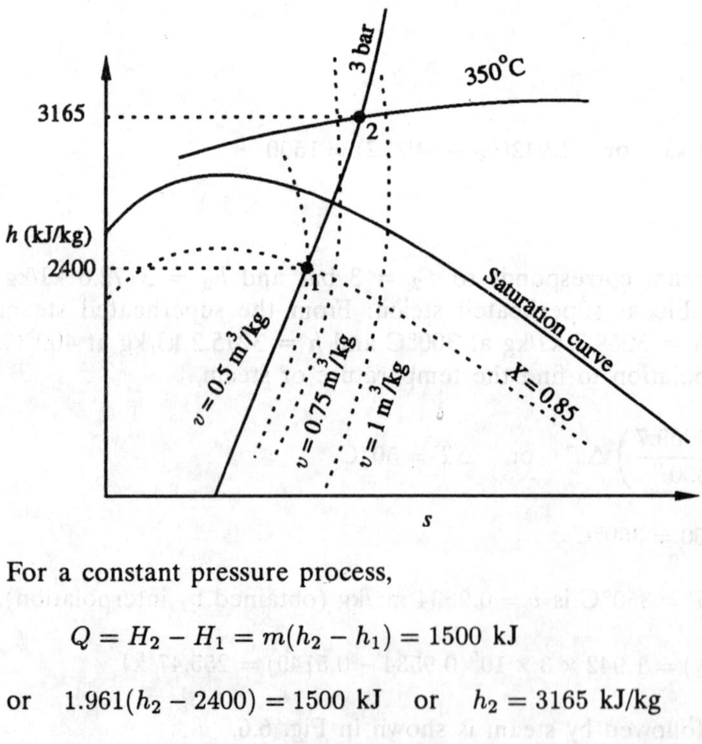

Fig. 6.7 Sketch for Example 6.4.

For a constant pressure process,

$$Q = H_2 - H_1 = m(h_2 - h_1) = 1500 \text{ kJ}$$

or $\quad 1.961(h_2 - 2400) = 1500$ kJ $\quad$ or $\quad h_2 = 3165$ kJ/kg

Locate state 2 at h_2 = 3165 kJ/kg and P = 3 bar as shown in Fig. 6.7. Then read v_2 = 0.96 m^3/kg (by interpolation between v = 0.75 m^3/kg and v = 1.0 m^3/kg curves.) Work done by steam,

$$W = mP(v_2 - v_1) = 1.961 \times 3 \times 10^5 \times (0.96 - 0.51) = 264.74 \text{ kJ}$$

EXAMPLE 6.5 A reversible and adiabatic steam turbine receives 1 kg/s of superheated steam at 10 bar and 300°C. The steam leaves the turbine at 1 bar. Neglecting the KE and PE changes, estimate the power output of the turbine. Use the Mollier diagram to solve the problem.

SOLUTION The First Law of Thermodynamics for a control volume operating under steady-state, steady-flow conditions is given by

$$\left(h_e + \frac{\mathbf{V}_e^2}{2} + gz_e\right) - \left(h_i + \frac{\mathbf{V}_i^2}{2} + gz_i\right) = \frac{\dot{Q} - \dot{W}_s}{\dot{m}} \tag{A}$$

If KE and PE changes are ignored and $\dot{Q}$ = 0 (adiabatic), Eq. (A) reduces to

$$\dot{W}_s = \dot{m}(h_i - h_e) \tag{B}$$

The Second Law of Thermodynamics for a reversible, adiabatic device gives

$$s_e = s_i \tag{C}$$

Locate the initial state i of steam at the intersection of P = 10 bar and T = 300°C curves as shown in Fig. 6.8.

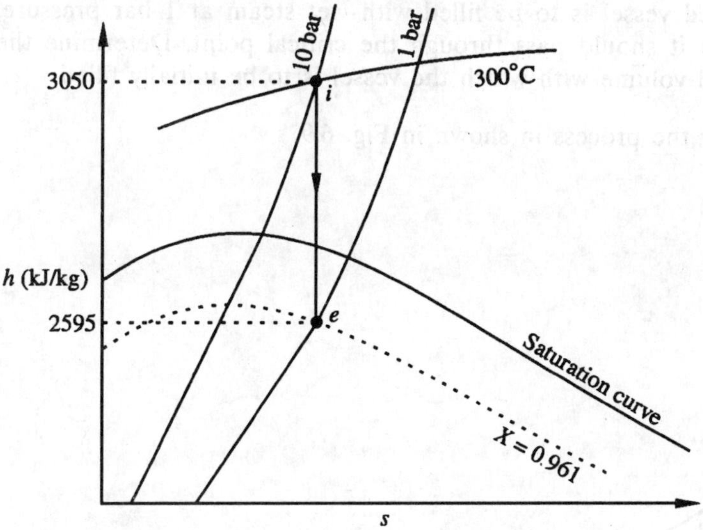

Fig. 6.8 *Sketch for Example 6.5.*

Since the process is isentropic, follow the constant entropy line (vertical line) from state i and locate the intersection point with P = 1 bar (shown as point e in Fig. 6.8). Then read

h_i = 3050 kJ/kg and h_e = 2595 kJ/kg from the Mollier diagram. Therefore, the power output of the turbine is given by (see Eq. (B)),

$$\dot{W}_s = \dot{m}(h_i - h_e) = 1 \times (3050 - 2595) \times 10^3 = 455 \text{ kW}$$

EXAMPLE 6.6 Rework Example 6.5 using the steam tables.

SOLUTION For steady-state, steady-flow conditions in a reversible adiabatic turbine, the first and second laws of thermodynamics give $\dot{W}_s = \dot{m}(h_i - h_e)$ and $s_e = s_i$. From the superheated steam tables, we read at P = 10 bar and T = 300°C, h = 3052.1 kJ/kg and s = 7.1251 kJ/kg K.

$$s_i = s_e = 7.1251 \text{ kJ/kg K}$$

From the saturated steam (pressure) tables, read the following values at 1 bar

$$s_f = 1.3027 \text{ kJ/kg K} \quad \text{and} \quad s_g = 7.3598 \text{ kJ/kg K}$$

$$h_f = 417.54 \text{ kJ/kg} \quad \text{and} \quad h_g = 2675.4 \text{ kJ/kg}$$

Since $s_f < s_e < s_g$, the final state is saturated steam of quality X_e.

Then $s_e = X_e s_g + (1 - X_e)s_f$

$$7.1251 = 7.3598X_e + 1.3027(1 - X_e) \quad \text{or} \quad X_e = 0.961$$

$$h_e = h_f(1 - X_e) + X_e h_g = 417.54(1 - 0.961) + 0.961 \times 2675.4 \quad \text{or} \quad h_e = 2587.3 \text{ kJ/kg}$$

Therefore, $\dot{W}_s = \dot{m}(h_i - h_e) = 1 \times (3052.1 - 2587.3) = 464.8$ kW

EXAMPLE 6.7 A rigid and closed vessel is to be filled with wet steam at 1 bar pressure, such that upon heating the system it should pass through the critical point. Determine the ratio of vapor volume to the liquid volume with which the vessel is to be initially filled.

SOLUTION The P–v diagram for the process in shown in Fig. 6.9.

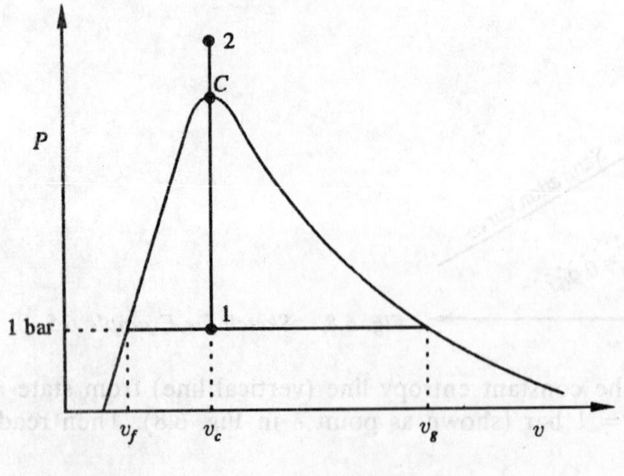

Fig. 6.9 Sketch for Example 6.7.

From the saturated steam (pressure) table, read the following values at $P = 1$ bar:

$$v_f = 0.001\ 043\ 4\ \text{m}^3/\text{kg} \quad \text{and} \quad v_g = 1.694\ \text{m}^3/\text{kg}.$$

At the critical point (221.2 bar, 374.15°C), $v_c = 0.003\ 17\ \text{m}^3/\text{kg}$

Let V = volume of the vessel, m = total mass of wet steam

Then initial specific volume, $v_1 = V/m$

Since the vessel is rigid (V = constant) and closed (m = constant), the specific volume of water remains constant during heating. The path followed by water is shown in Fig. 6.9 as 1-C-2, which passes through the critical point (C). Therefore,

$$v_2 = v_c = v_1 \quad \text{or} \quad v_c = 0.003\ 17 = v_1 = (1 - X_1)v_f + XV_g$$

or $\quad 0.003\ 17 = (1 - X_1)0.001\ 043\ 4 + 1.694X_1 \quad$ or $\quad X_1 = 1.256 \times 10^{-3}$

$$\frac{\text{volume of vapor}}{\text{volume of liquid}} = \frac{m_g v_g}{m_f v_f} = \frac{(m_g v_g/m)}{(m_f v_f/m)} = \frac{X_1 v_g}{(1 - X_1)v_f} = \frac{1.256 \times 10^{-3} \times 1.694}{(1 - 1.256 \times 10^{-3}) \times 0.001\,043\,4} = 2.04$$

EXAMPLE 6.8 At a particular instant, a steam radiator of volume 0.1 m^3 is filled with steam at 1.5 bar and 220°C. The inlet and outlet valves of the radiator are closed. After some time, the steam pressure in the radiator was found to be 1 bar, due to heat losses to the surroundings. Determine the masses of vapor and liquid water in the final state. Make use of the Mollier diagram to solve the problem and sketch the process followed by the steam on a P–v diagram.

SOLUTION A schematic of the Mollier diagram showing the initial and final states of the steam is presented in Fig. 6.10.

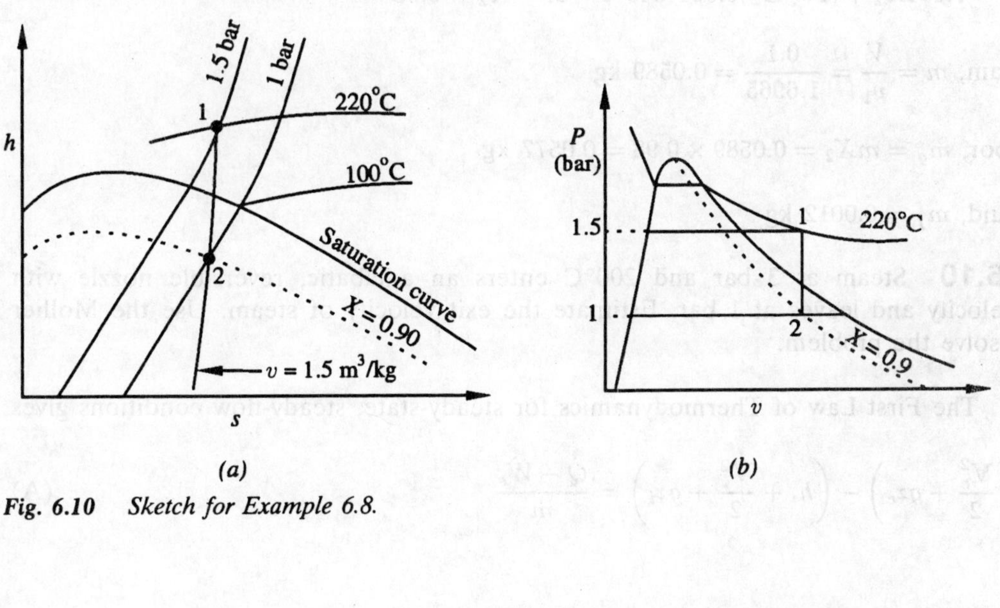

Fig. 6.10 Sketch for Example 6.8.

Locate the initial state 1 at the intersection of T = 220°C and P = 1.5 bar curves. At this state, read v_1 = 1.5 m^3/kg. The steam follows the constant volume process. Hence, now locate the final state 2 at the intersection of P = 1 bar and v = 1.5 m^3/kg curves. Then read X = 0.9 at state 2. Mass of steam, m = 0.1/0.5 = 0.067 kg. Mass of vapor in the final state, $m_g = mX = 0.067 \times 0.9$ = 0.06 kg. Mass of liquid in the final state, m_f = 0.007 kg. The P–v diagram is shown in Fig. 6.10(b).

EXAMPLE 6.9 Rework Example 6.8 using the steam tables.

SOLUTION Initial state of steam is P = 1.5 bar and T = 220°C. From the superheated steam tables read the following values:

At P = 1 bar: v = 2.1720 m^3/kg at 200°C and v = 2.6390 m^3/kg at 300°C. By interpolation, we get at P = 1 bar and T = 220°C, v = 2.2654 m^3/kg.

At P = 2 bar: v = 1.0804 m^3/kg at 200°C and v = 1.3162 m^3/kg at 300°C. By interpolation, we get at P = 2 bar and T = 220°C, v = 1.1276 m^3/kg.

Interpolating between the above two values, we get v = 1.6965 m^3/kg at P = 1.5 bar and T = 220°C.

The volume of the radiator and the mass of steam are constant. Hence, the specific volume (V/m) of steam is constant during cooling. Therefore, $v_2 = v_1$ = 1.6965 m^3/kg.

From saturated steam (pressure) table, we find at P = 1 bar:

$$v_f = 0.001\,043\,4 \text{ m}^3/\text{kg} \quad \text{and} \quad v_g = 1.694 \text{ m}^3/\text{kg}$$

$$v_f < v_2 < v_g$$

Hence, the final state is wet steam of quality X_2.

$$v_2 = X_2 v_g + (1 - X_2) v_f$$

$$1.6965 = 1.694 X_2 + (1 - X_2) 0.001\,043.4 \quad \text{or} \quad X_2 = 0.98$$

$$\text{Mass of steam, } m = \frac{V}{v_1} = \frac{0.1}{1.6965} = 0.0589 \text{ kg}$$

$$\text{Mass of vapor, } m_g = mX_2 = 0.0589 \times 0.98 = 0.0577 \text{ kg}$$

$$\text{Mass of liquid, } m_f = 0.0012 \text{ kg}$$

EXAMPLE 6.10 Steam at 3 bar and 200°C enters an adiabatic, reversible nozzle with negligible velocity and leaves at 1 bar. Estimate the exit velocity of steam. Use the Mollier diagram to solve the problem.

SOLUTION The First Law of Thermodynamics for steady-state, steady-flow conditions gives

$$\left(h_e + \frac{\mathbf{V}_e^2}{2} + gz_e\right) - \left(h_i + \frac{\mathbf{V}_i^2}{2} + gz_i\right) = \frac{\dot{Q} - \dot{W}_s}{\dot{m}} \qquad \text{(A)}$$

For an adiabatic nozzle, $\dot{Q} = 0$ and $\dot{W}_s = 0$. Inlet velocity is negligible ($\mathbf{V}_i = 0$). Ignore changes in PE. Then Eq. (A) reduces to

$$\mathbf{V}_e = \sqrt{2(h_i - h_e)} \tag{B}$$

For a reversible and adiabatic process, the Second Law of Thermodynamics gives

$$s_e = s_i \tag{C}$$

A sketch of the Mollier diagram is shown in Fig. 6.11.

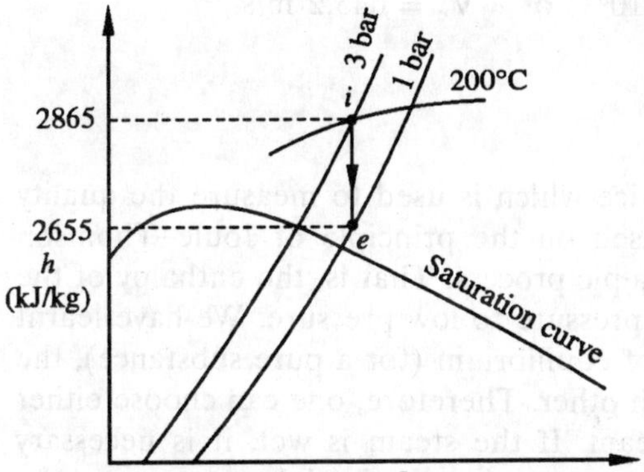

***Fig. 6.11** Sketch for Example 6.10.*

Locate the initial state i at the intersection of the curves P = 3 bar and T = 200°C as shown in Fig. 6.11. Since the process is isentropic, follow the constant entropy line (vertical line) and locate the final state e at the intersection of the vertical line from state i and $P = 1$ bar curve. Then read h_i = 2865 kJ/kg and h_e = 2655 kJ/kg.

$$\mathbf{V}_e = \sqrt{2(h_i - h_e)} = \sqrt{2(2865 - 2655) \times 10^3} = 648 \text{ m/s}$$

EXAMPLE 6.11 Rework Example 6.10 using the steam tables.

SOLUTION For a reversible, adiabatic nozzle (if the inlet velocity is ignored), we get (see Example 6.10)

$$\mathbf{V}_e = \sqrt{2(h_i - h_e)} \quad \text{and} \quad s_e = s_i$$

From the superheated steam tables, obtain the following values at $P = 3$ bar and T = 200°C:

$$h_i = 2865.5 \text{ kJ/kg} \quad \text{and} \quad s_i = 7.3119 \text{ kJ/kg K}$$

$$s_e = s_i = 7.3119 \text{ kJ/kg K}$$

The value of $s_e < s_g$ (= 7.3598 kJ/kg K) at 1 bar. Hence, the exit steam is wet. Then from the saturated steam tables at 1 bar we read,

$$h_f = 417.54 \text{ kJ/kg} \quad \text{and} \quad s_f = 1.3027 \text{ kJ/kg K}$$

$$h_g = 2675.4 \text{ kJ/kg} \quad \text{and} \quad s_g = 7.3598 \text{ kJ/kg K}$$

$$s_e = X_e s_g + (1 - X_e)s_f$$

$$7.3119 = 7.3598X_e + (1 - X_e)1.3027 \quad \text{or} \quad X_e = 0.992$$

$$h_e = 2675.4 \times 0.992 + (1 - 0.992)417.54 = 2657.34 \text{ kJ/kg}$$

Therefore,

$$\mathrm{V}_e = \sqrt{2(h_i - h_e)} = \sqrt{2 \times (2865.5 - 2657.34) \times 10^3} \quad \text{or} \quad \mathrm{V}_e = 645.2 \text{ m/s}$$

6.7 Throttling Calorimeter

A throttling calorimeter (shown in Fig. 6.12) is a device which is used to measure the quality or dryness fraction of steam. This calorimeter is based on the principle of Joule–Thomson expansion. We have learnt that throttling is an isenthalpic process. That is, the enthalpy of the fluid remains constant when it is throttled from high pressure to low pressure. We have learnt that when liquid and vapor phases coexist in a state of equilibrium (for a pure substance), the temperature and pressure are not independent of each other. Therefore, one can choose either pressure or temperature to specify the saturated steam. If the steam is wet, it is necessary to specify its quality also. The quality of wet steam cannot be obtained by measuring the temperature and pressure of wet steam, because these two variables are not independent of each other. To determine the quality of wet steam, it can be throttled to low pressure, such that the exit state is superheated. Then one can measure the temperature and pressure of the superheated steam. The pressure and temperature of the superheated steam being known, its enthalpy can be determined. Then

$$h_e = h_i = h_f + X_i h_{fg} \tag{6.8}$$

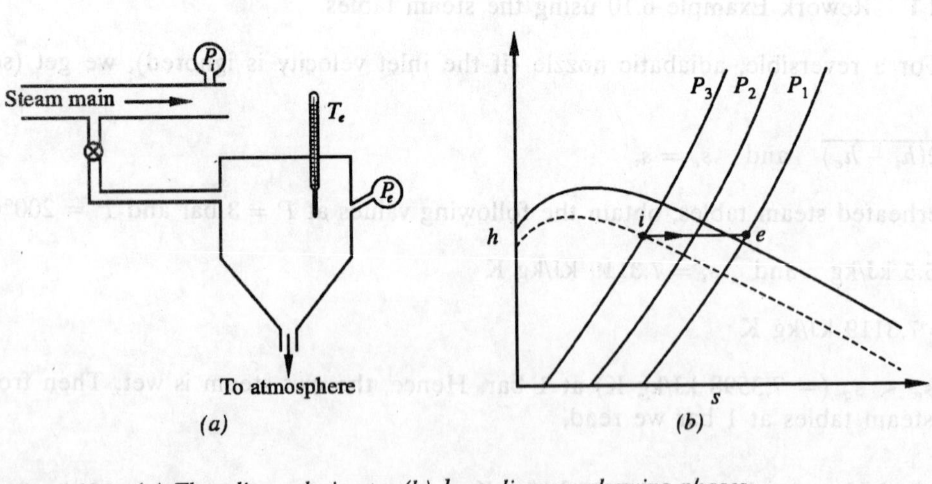

Fig. 6.12 (a) Throttling calorimeter (b) h–s diagram showing phases.

With the value of either temperature or pressure of the steam entering the throttling calorimeter, the enthalpies of saturated liquid and saturated vapor can be obtained from the steam tables. Then the quality or dryness fraction X_i of the wet steam can be determined by using Eq. (6.8).

EXAMPLE 6.12 A throttling calorimeter is used to measure the quality of wet steam at 10 bar flowing through a pipeline. The pressure and temperature at the exit of the calorimeter are measured as 1 bar and 110°C. Determine the quality of steam.

SOLUTION Throttling is an isenthalpic process. Hence, $h_1 = h_2$.

For superheated steam at 1 bar and 110°C, h_2 = 2696.1 kJ/kg (obtained by interpolation).

From the saturated steam tables, we read at P = 10 bar, h_f = 762.61 kJ/kg and h_g = 2776.2 kJ/kg.

Therefore,

$$h_2 = h_g X_1 + (1 - X_1) h_f$$

$$2696.1 = 2776.2 X_1 + (1 - X_1) 762.61 \quad \text{or} \quad X_1 = 0.96$$

EXAMPLE 6.13 Rework Example 6.12 using the Mollier diagram.

SOLUTION A schematic of the Mollier diagram indicating the final and initial states of steam is shown in Fig. 6.13.

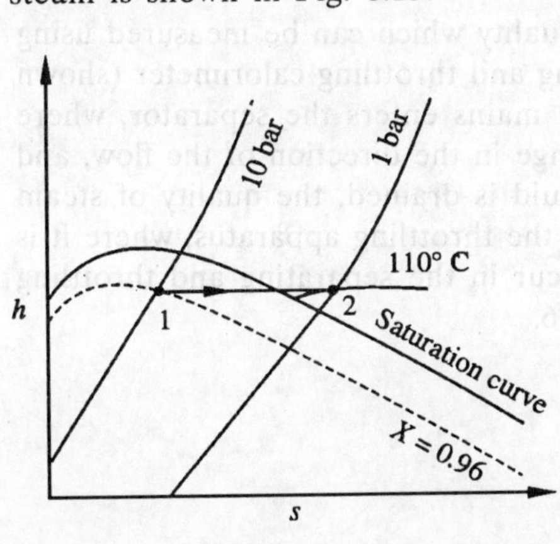

Fig. 6.13 Sketch for Example 6.13.

Locate state 2 at the intersection of P = 1 bar and T = 110°C curves. Then follow the constant enthalpy line (horizontal line) and locate its intersection with the P = 10 bar curve in the two-phase region (state 1). At this point, read X_1 = 0.96. Hence, quality of steam = 0.96.

6.8 Combined Separating and Throttling Calorimeter

Suppose the quality of steam at 10 bar is 0.9. Unless this steam is throttled to a pressure lower than 0.07 bar, the exit steam will not be superheated. Usually the steam has to be throttled to a low pressure such that its temperature has to be at least 5°C more than the saturation temperature. Moreover, the exit pressure of a throttling calorimeter is the atmospheric pressure. If the steam leaving the throttling calorimeter is at 1 bar and 105°C, then the minimum quality of the steam which can be measured by a throttling calorimeter depends on the pressure of the wet steam as shown in Fig. 6.14.

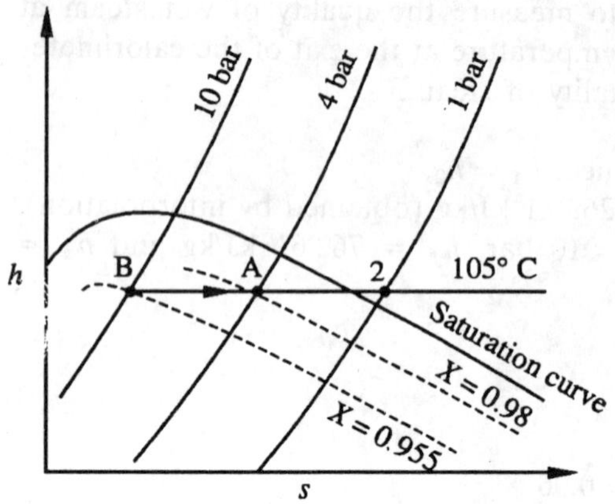

Fig. 6.14 Mollier diagram indicating the minimum quality which can be measured with a throttling calorimeter.

If the wet steam is at 4 bar, then the state of the steam corresponds to point A, corresponding to $X = 0.98$. On the other hand, if the wet steam is at 10 bar, then the state of steam corresponds to point B with a quality of 0.955.

If the quality of wet steam is less than the minimum quality which can be measured using a simple throttling calorimeter, then a combined separating and throttling calorimeter (shown in Fig. 6.15) is used. In this calorimeter, steam from the mains enters the separator, where part of the moisture separates out due to the sudden change in the direction of the flow, and the separated liquid is drained out. When part of the liquid is drained, the quality of steam leaving the separator will be high. Then this steam enters the throttling apparatus, where it is throttled to the superheated state. The processes that occur in the separating and throttling calorimeter are shown on the Mollier diagram in Fig. 6.16.

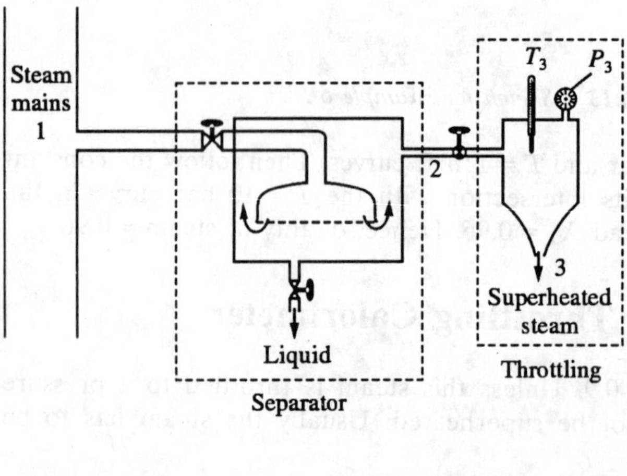

Fig. 6.15 Schematic of a combined separating and throttling calorimeter.

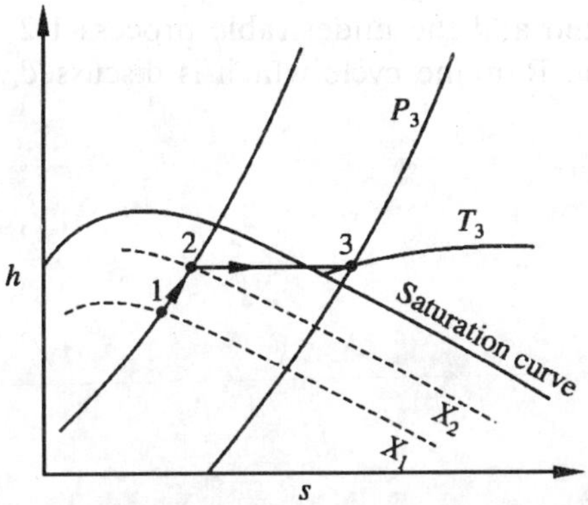

Fig. 6.16 *Mollier diagram showing the processes that occur in a separating and throttling calorimeter.*

In the separator, the path 1-2 is the constant pressure process. If m_s kg of liquid is separated in the separator from m kg of wet steam at state 1, then $(m - m_s)$ kg at state 2 is throttled. If X_2 is the quality of $(m - m_s)$ kg of steam, then the mass of dry vapor in that steam is $(m - m_s)X_2$. Hence the quality of the original wet steam is given by

$$X_1 = \frac{(m - m_s)X_2}{m} \tag{6.9}$$

6.9 Carnot Vapor Power Cycle

Historically, man's endeavor for continuous conversion of heat into work led to the development of thermodynamics. Hence, it is natural that an engineer should be familiar with the heat engine cycles or power cycles. Different types of working fluids are employed in the power cycles. The nature of the working fluids can be classified into two groups as:

1. Vapor power cycles in which the working fluid undergoes a phase change during the cyclic process, and
2. Gas power cycle in which the working fluid does not undergo any phase change.

The Carnot cycle which consists of only reversible processes has the highest possible efficiency. Therefore, a study of the Carnot cycle employing vapor as a working medium provides a good basis for the study of vapor power cycles. Consider the Carnot cycle shown in Fig. 6.17. Wet vapor at state 1 is isentropically compressed to state 2, where it is available as saturated liquid. The saturated liquid at state 2 undergoes a phase change at constant temperature by absorbing energy from a thermal reservoir till it is completely converted into dry saturated vapor at state 3. Then the vapor at state 3 undergoes reversible, adiabatic expansion to state 4. After which the wet vapor at state 4 enters a condenser where it rejects energy at constant temperature till it reaches state 1. Thus the vapor undergoes a cyclic change.

Though it is quite desirable to operate a power plant on the Carnot cycle, it is not practical because of the difficulties associated with isentropic compression of the wet vapor from state 1 to state 2. Moreover, the rate of delivery of work by such a plant will be so low that it is not worthwhile to pursue it. The desirable and practical reversible processes (2-3 and 4-1)—energy addition and rejection at constant temperature—are retained and the undesirable process 1-2 is replaced by a technically feasible process, by devising the Rankine cycle which is discussed below.

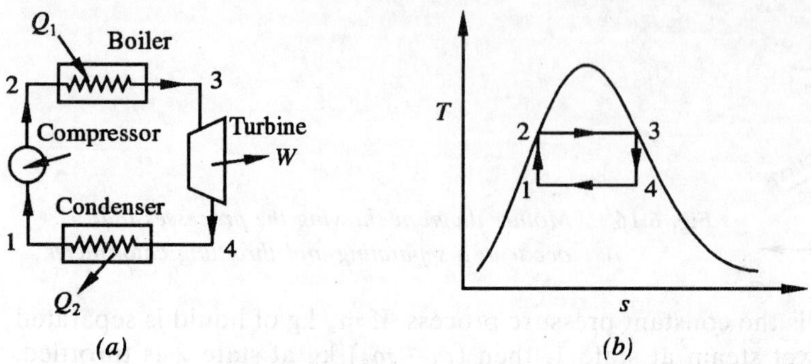

Fig. 6.17 (a) Schematic diagram of a Carnot engine (b) T–s diagram of a Carnot vapor power cycle.

6.10 Ideal Rankine Cycle

Professor W J Rankine (1820–1872) modified the Carnot cycle and presented a technically feasible cycle. In this modification, the isentropic compression of wet vapor is replaced by the isentropic compression of saturated liquid. The Rankine cycle and its modifications are used in thermal power plants and the working fluid is water. The ideal Rankine cycle is shown in Fig. 6.18.

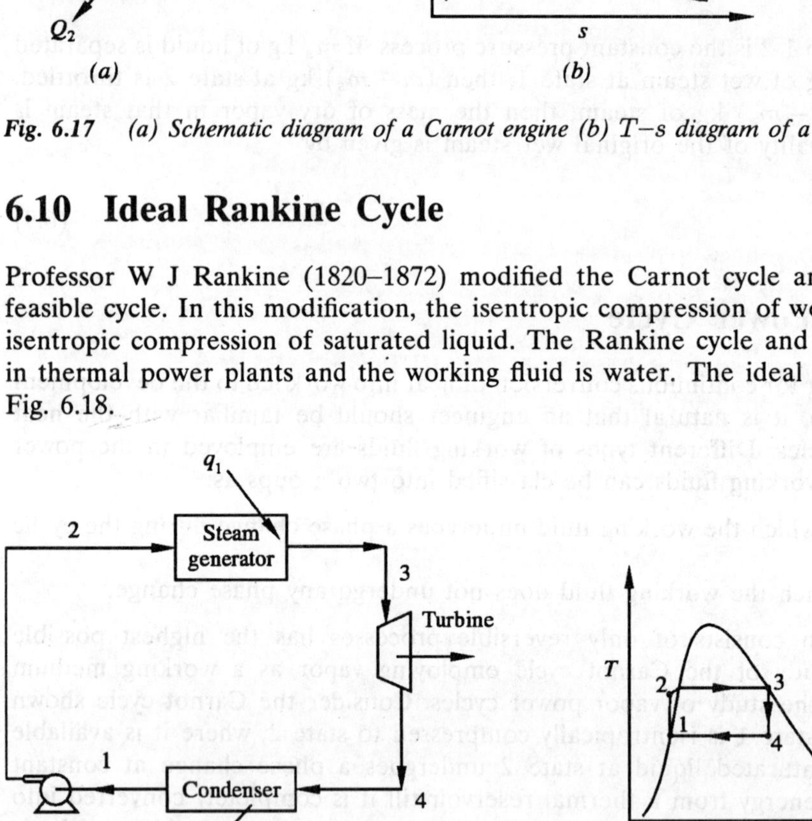

Fig. 6.18 (a) Schematic representation of an ideal Rankine cycle (b) T–s diagram of an ideal Rankine cycle.

In the simple Rankine cycle, saturated liquid water at low pressure (state 1) leaving the condenser is isentropically pumped into the steam generator, which operates at high pressure (state 2). Energy addition as heat (q_1) takes place in the steam generator and the steam leaves at state 3. The steam at state 3 enters a reversible and adiabatic turbine and expands to a low pressure (state 4). The steam at low pressure passes through a condenser, where it rejects energy (q_2) as heat by cooling into water and leaving as a saturated liquid (state 1). Application of the First Law of Thermodynamics to the control volumes (pump, steam generator, turbine and condenser), gives

Work done on pump, per kg of water, $W_p = h_2 - h_1$

Energy added in steam generator, $q_1 = h_3 - h_2$

Work delivered by turbine, $W_T = h_3 - h_4$

Energy rejected in the condenser, $q_2 = h_4 - h_1$

Therefore, the thermal efficiency of the Rankine cycle is given by

$$\eta = \frac{q_1 - q_2}{q_1} = \frac{(h_3 - h_2) - (h_4 - h_1)}{h_3 - h_2} = \frac{(h_3 - h_4) - (h_2 - h_1)}{h_3 - h_2} = \frac{\text{Net work done}}{\text{Energy absorbed}} \quad (6.10)$$

Practical Rankine Cycle In actual practice, the pump and turbine have certain irreversibilities. Therefore, the processes 1-2 and 3-4 are not isentropic. The Second Law of Thermodynamics tells us that the exit entropy of a fluid is greater than the inlet entropy, if there is any irreversibility in the device. Hence,

$$s_2' > s_2$$

$$s_4' > s_4$$

where s_2' and s_4' represent the entropy of the fluid leaving the actual pump and actual turbine, respectively. Moreover, if dry saturated steam enters a turbine and undergoes expansion to a low pressure, the quality of steam keeps decreasing during the expansion. If the quality of steam in a turbine is less than 0.9, the presence of water droplets in the steam erode the turbine blades. Therefore, it is necessary to maintain a high quality of steam in the turbine. The practical Rankine cycle is shown in Fig. 6.19.

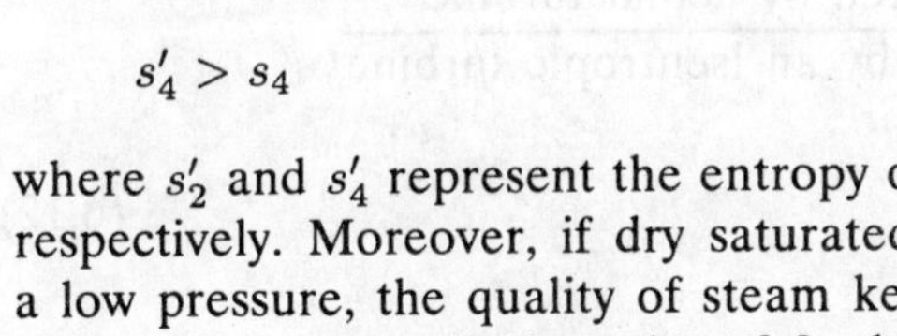

Fig. 6.19 T–s diagram of a practical Rankine cycle.

In Fig. 6.19, 1-2-A-B-1 represents a simple Rankine cycle, in which dry saturated steam enters the turbine at state A and leaves at state B. The quality of steam leaving the turbine is X_B. In the Rankine cycle 1-2-3-4-1, the pressure of steam leaving the steam generator is the same as that of the cycle 1-2-A-B-1, but now the steam is superheated to state 3. The superheated steam undergoes isentropic expansion to state 4 and the quality of steam leaving the turbine X_4 is such that $X_4 > X_B$. The practical Rankine cycle is shown as 1-2′-3-4′-1, in which the pump and turbine are not isentropic.

In the actual turbine, the work delivered is less than the isentropic turbine. Similarly, the work consumed by an actual pump is greater than the work consumed by an isentropic pump. That is,

$$h_3 - h_4' < h_3 - h_4$$

$$h_2' - h_1 > h_2 - h_1$$

The thermal efficiency of a practical Rankine cycle is given by

$$\eta = \frac{\text{Net work}}{\text{Energy absorbed}} = \frac{(h_3 - h_4') - (h_2' - h_1)}{h_3 - h_2'} \tag{6.11}$$

The performance of an actual turbine or pump is usually expressed in terms of isentropic efficiency. They are defined as

$$\text{Isentropic efficiency of turbine, } \eta_T = \frac{\text{Work delivered by actual turbine}}{\text{Work delivered by an isentropic turbine}}$$

$$\eta_T = \frac{h_3 - h_4'}{h_3 - h_4} \tag{6.12}$$

$$\text{Isentropic efficiency of pump, } \eta_P = \frac{\text{Work required by isentropic pump}}{\text{Work required by actual pump}}$$

$$\eta_P = \frac{h_2 - h_1}{h_2' - h_1} \tag{6.13}$$

EXAMPLE 6.14 In a thermal power plant operating on an ideal Rankine cycle, superheated steam at 30 bar and 400°C enters a turbine and leaves at 0.1 bar. The steam at low pressure is condensed and pumped into the steam generator. Assuming that the turbine and pump are isentropic, estimate the thermal efficiency of the power plant.

SOLUTION For steam at 30 bar and 400°C, read the following values from the superheated steam tables.

$$s_3 = 6.9246 \text{ kJ/kg K and } h_3 = 3232.5 \text{ kJ/kg}$$

For steam at 0.1 bar, read the following values from the saturated steam (pressure) tables.

$h_f = 191.83$ kJ/kg; $h_g = 2584.8$ kJ/kg

$v_f = 0.001\ 010\ 2$ m^3/kg; $s_f = 0.6493$ kJ/kg K and $s_g = 8.1511$ kJ/kg K

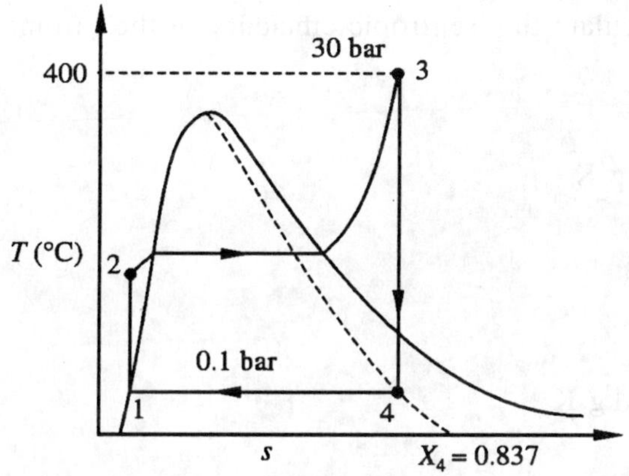

Fig. 6.20 *Sketch for Example 6.14.*

The turbine is isentropic. Therefore, $s_3 = s_4$

$$s_4 = (1 - X_4)s_f + s_g X_4$$

$$6.9246 = (1 - X_4)0.6493 + 8.1511X_4 \quad \text{or} \quad X_4 = 0.837$$

Then

$$h_4 = h_f(1 - X_4) + X_4 h_g = 191.83(1 - 0.837) + 0.837 \times 2584.8 = 2194.75 \text{ kJ/kg}$$

$$h_1 = h_f = 191.83 \text{ kJ/kg}$$

We know that

$$dh = d(u + Pv) = du + Pdv + vdP \quad \text{and} \quad du = Tds - Pdv$$

Therefore, $dh = Tds + vdP$

For an isentropic pump

$$h_2 - h_1 = \int_1^2 dh = \int_1^2 vdP = v_f(P_2 - P_1)$$

Since the liquid is incompressible,

$$h_2 - h_1 = 0.001\ 010\ 2(30 - 0.1) \times 10^5 = 3.02 \text{ kJ/kg}$$

$$h_2 = h_1 + 3.02 = 191.83 + 3.02 = 194.85 \text{ kJ/kg}$$

Therefore, thermal efficiency is given by

$$\eta = \frac{(h_3 - h_4) - (h_2 - h_1)}{h_3 - h_2} = \frac{(3232.5 - 2194.75) - 3.02}{3232.5 - 194.85} \quad \text{or} \quad \eta = 0.34$$

EXAMPLE 6.15 In a thermal power plant, steam at 40 bar and 400°C enters a turbine and leaves as wet steam of quality 0.9 at 0.2 bar. Calculate the isentropic efficiency of the turbine.

SOLUTION Steam at 40 bar and 400°C:

$$h_i = 3215.7 \text{ kJ/kg} \quad \text{and} \quad s_i = 6.7733 \text{ kJ/kg K}$$

Steam at 0.2 bar:

$$h_f = 251.45 \text{ kJ/kg}; \quad h_g = 2609.9 \text{ kJ/kg}$$

$$s_f = 0.8321 \text{ kJ/kg K} \quad \text{and} \quad s_g = 7.9094 \text{ kJ/kg K}$$

$$s'_e = s_f(1 - X'_e) + X'_e s_g$$

$$s'_e = 0.8321(1 - 0.9) + 0.9 \times 7.9094 = 7.2017 \text{ kJ/kg K}$$

Since $s'_e > s_i$, the turbine is nonisentropic.

$$h'_e = (1 - X'_e)h_f + X'_e h_g = (1 - 0.9)251.45 + 0.9 \times 2609.9 = 2374.06 \text{ kJ/kg}$$

For an isentropic turbine

$$s_i = s_e = s_f(1 - X_e) + X_e s_g$$

$$6.7733 = 0.8321(1 - X_e) + 7.9094X_e \quad \text{or} \quad X_e = 0.839$$

$$h_e = h_f(1 - X_e) + X_e h_g = 251.45(1 - 0.839) + 0.839 \times 2609.9 = 2230.19 \text{ kJ/kg}$$

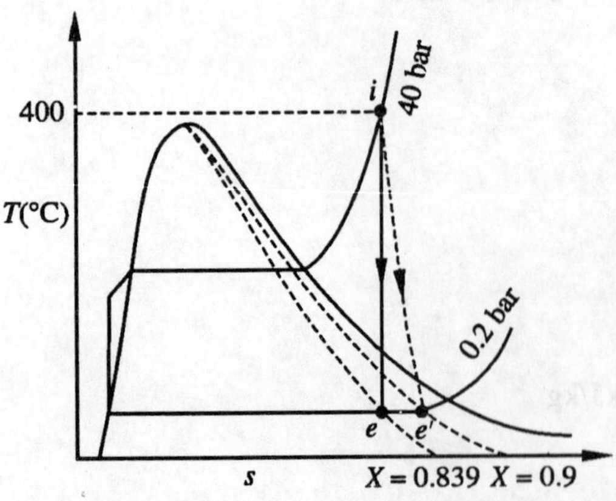

Fig. 6.21 Sketch for Example 6.15.

In Fig. 6.21, the path i-e represents the process for an isentropic turbine and the path i-e' denotes the process in an actual turbine.

Isentropic efficiency of the turbine is given by

$$\eta_T = \frac{h_i - h_e'}{h_i - h_e} = \frac{3215.7 - 2374.06}{3215.7 - 2230.19} = 0.854$$

EXAMPLE 6.16 Rework Example 6.14 if the isentropic efficiency of the turbine is 0.8.

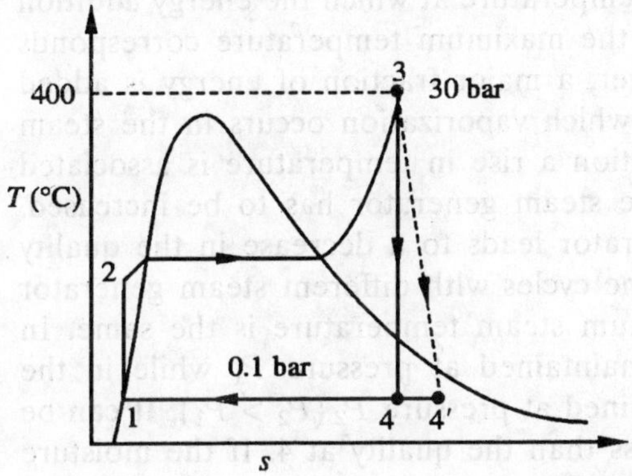

Fig. 6.22 Sketch for Example 6.16.

SOLUTION For steam at 30 bar and 400°C:

$$h_3 = 3232.5 \text{ kJ/kg} \quad \text{and} \quad s_3 = 6.9246 \text{ kJ/kg K}$$

For steam at 0.1 bar:

$$h_f = 191.83 \text{ kJ/kg}; \quad h_g = 2584.8 \text{ kJ/kg}; \quad v_f = 0.001\ 010\ 2 \text{ m}^3\text{/kg}$$

$$s_f = 0.6493 \text{ kJ/kg K} \quad \text{and} \quad s_g = 8.1511 \text{ kJ/kg K}$$

For an isentropic turbine: $X_4 = 0.837$ and $h_4 = 2194.75$ kJ/kg

Work done on pump: $h_2 - h_1 = 3.02$ kJ/kg; $h_2 = 194.85$ kJ/kg (see Example 6.14)

Isentropic efficiency of the turbine is given by

$$\eta_T = \frac{h_3 - h_4'}{h_3 - h_4}$$

$$0.8 = \frac{3232.5 - h_4'}{3232.5 - 2194.75} \quad \text{or} \quad h_4' = 2402.3 \text{ kJ/kg}$$

Thermal efficiency is given by

$$\eta = \frac{(h_3 - h_4') - (h_2 - h_1)}{h_3 - h_2} = \frac{(3232.5 - 2402.3) - 3.02}{3232.5 - 194.85} = 0.272$$

6.11 Reheat Rankine Cycle

The thermal efficiency of any power cycle can be increased either by increasing the temperature at which the energy addition takes place while keeping the sink temperature constant, or by decreasing the sink temperature while the temperature at which the energy addition takes place is held constant.

The sink temperature is determined by the environment or the availability of cooling water. Any effort to reduce the sink temperature requires the performance of work and hence not economical. Therefore, to increase the efficiency, the temperature at which the energy addition takes place has to be increased. In a Rankine cycle, the maximum temperature corresponds to the temperature of the superheated steam. However, a major fraction of energy is added during phase change. Therefore, the temperature at which vaporization occurs in the steam generator should be increased. Since during vaporization a rise in temperature is associated with a rise in pressure, the operating pressure of the steam generator has to be increased. An increase in operating pressure of the steam generator leads to a decrease in the quality of steam leaving the turbine. In Fig. 6.23, two Rankine cycles with different steam generator pressures are shown. For both the cycles, the maximum steam temperature is the same. In the Rankine cycle 1-2-3-4, the steam generator is maintained at pressure P_1 while in the Rankine cycle 1-2′-3′-4′ the steam generator is maintained at pressure P_2 ($P_2 > P_1$). It can be clearly seen from Fig. 6.23 that the quality at 4′ is less than the quality at 4. If the moisture content in the turbine exceeds 10%, it causes serious erosion of the turbine blades and hence, operating the steam generator at high pressure is not desirable.

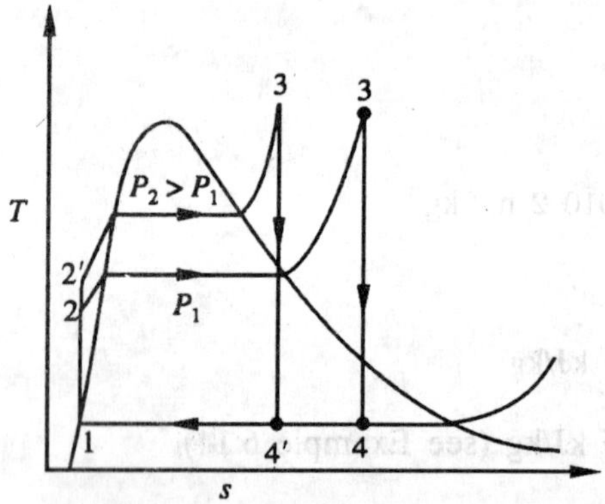

Fig. 6.23 *Rankine cycles with different steam generator pressures.*

The reheat modification of the Rankine cycle (or the reheat Rankine cycle) was developed to take advantage of increased efficiency by increasing the average temperature at which the energy addition takes place and at the same time avoids the high moisture content in the turbine. The reheat cycle is shown in Fig. 6.24.

In the reheat Rankine cycle, superheated steam at state 3 enters the turbine and expands to some intermediate pressure only (state 4). Then the steam is withdrawn from the turbine at state 4 and fed to the superheater of the steam generator. The steam is reheated to state 5 before it is supplied to the turbine for further expansion. Now the steam expands in the turbine to state 6, before it passes through the condenser. The energy added (per unit mass of steam) in the steam generator is given by

$$q_1 = (h_3 - h_2) + (h_5 - h_4)$$

The energy rejected in the condenser, $q_2 = h_6 - h_1$. Therefore, the thermal efficiency is given by

$$\eta = \frac{(h_3 - h_2) + (h_5 - h_4) - (h_6 - h_1)}{(h_3 - h_2) + (h_5 - h_4)} \tag{6.14}$$

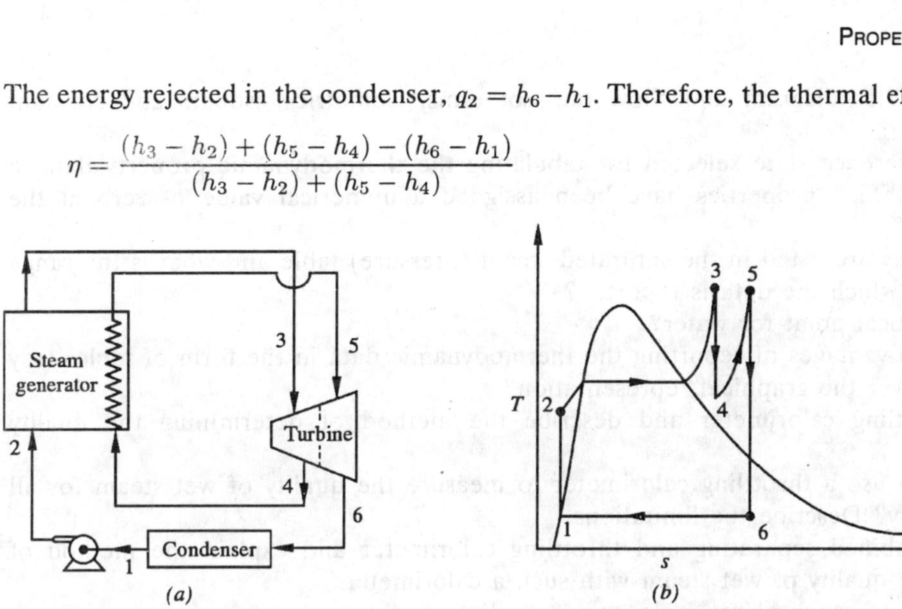

Fig. 6.24 *(a) Schematic representation of a reheat Rankine cycle (b) T–s diagram of a reheat Rankine cycle.*

Review Questions

1. Define isothermal compressibility.
2. In how many states can water be found and what are they?
3. What is a pure substance and what is a compressible substance?
4. What is a saturated state?
5. What is a saturated liquid?
6. What is a saturated vapor?
7. What is a compressed liquid or subcooled liquid?
8. Suppose liquid water at 20°C is heated at constant pressure till it becomes superheated vapor. Sketch the process on a T–v diagram.
9. What is superheated vapor?
10. What is meant by saturation pressure and saturation temperature?
11. Sketch the temperature–volume diagram and show the constant pressure lines, liquid zone, vapor zone and two-phase region on the diagram.
12. Define the quality of a wet vapor.
13. Sketch the pressure versus volume diagram for a pure substance and indicate the saturated liquid curve, the saturated vapor curve, the constant quality lines and a few isotherms on the diagram.
14. Sketch the temperature versus entropy diagram for a pure substance and show the constant quality lines and constant pressure lines on the diagram.
15. Suppose a rigid and closed vessel is filled with a small quantity of saturated liquid, the remaining space is filled with a large quantity of saturated vapor and the vessel is heated. Sketch the path followed by the liquid on a P–v diagram.
16. Suppose a rigid and closed vessel is filled with a wet vapor of high quality and heated. Sketch the path followed by the wet vapor on a P–v diagram.

17. Sketch the Mollier diagram and show the isotherms, isometrics, isobars and constant quality lines.
18. What is the reference state selected for tabulating the thermodynamic property data in steam tables? Which properties have been assigned a numerical value of zero at the reference state?
19. Which properties are listed in the saturated steam (pressure) table and what is the range of pressure for which the data is reported?
20. What is the critical point for water?
21. What are the advantages of reporting the thermodynamic data in the form of tables (say steam tables) over the graphical representation?
22. Sketch a throttling calorimeter and describe the method of determining the quality of wet steam.
23. Is it possible to use a throttling calorimeter to measure the quality of wet steam for all ranges of quality? Describe the limitations.
24. Sketch the combined separating and throttling calorimeter and explain the method of determining the quality of wet steam with such a calorimeter.
25. Sketch the Carnot vapor power cycle on a T–s diagram.
26. Why is a Carnot vapor power cycle not practical?
27. What modifications are made in the Carnot vapor cycle while devising a Rankine cycle?
28. Sketch the simple Rankine cycle on a T–s diagram.
29. Sketch a practical Rankine cycle on a T–s diagram and explain how it differs from the simple Rankine cycle.
30. Define the isentropic efficiency of a turbine and a pump.
31. Is it possible to improve the efficiency of a Rankine cycle? Explain how.
32. Sketch two Rankine cycles—one operating with a steam generator at pressure P_2 ($P_2 > P_1$) and the other operating with a steam generator at pressure P_1—and explain the drawbacks associated with operating a steam generator at high pressure.
33. Sketch the reheat Rankine cycle on a T–s diagram.

Problems

6.1 A tank of volume 10 m^3 contains dry saturated steam at 200°C. Estimate the mass of steam in the tank.

6.2 A tank of volume 10 m^3 is filled with wet steam of quality 0.8 at 200°C. Estimate the masses of liquid and vapor in the tank.

6.3 It is required to store 100 kg of wet steam of quality 0.85 at 3 bar pressure. Estimate the volume of the tank.

6.4 Calculate the specific enthalpy and specific entropy of wet steam of quality 0.85 at 3 bar.

6.5 Calculate the specific internal energy of wet steam of quality 0.85 at 3 bar pressure.

6.6 A rigid vessel of volume 0.1 m^3 contains dry saturated water vapor at 2 bar. Owing to heat losses, if the pressure reduces to 1 bar, estimate the energy lost by steam.

6.7 A rigid vessel of volume 0.3 m^3 is filled with 0.1 m^3 of saturated liquid water at 3 bar, and the remaining space is filled with saturated water vapor at the same pressure. Now energy is transferred as heat till the vessel contains only dry saturated vapor. Estimate the final pressure of steam in the vessel and the quantity of energy added as heat.

6.8 A piston-cylinder assembly contains 0.2 kg of wet steam of quality 0.8 at 3 bar. Determine the final state of steam and the work done, if 300 kJ of energy is added as heat at constant pressure.

6.9 Rework Problem 6.8 using the Mollier diagram.

6.10 A piston-cylinder assembly contains 0.4 kg of superheated steam at 5 bar and 400°C. While keeping the pressure constant, 400 kJ of energy is withdrawn as heat. Determine the final state of the steam.

6.11 Rework Problem 6.10 using the Mollier diagram.

6.12 A piston-cylinder assembly contains 0.1 m^3 of superheated steam at 10 bar and 400°C. If the steam is allowed to expand reversibly and adiabatically to a pressure of 3 bar, calculate the work done by the steam. Use steam tables to solve the problem.

6.13 Rework Problem 6.12 the using the Mollier diagram.

6.14 In a thermal power plant, superheated steam at 30 bar and 400°C enters a reversible and adiabatic turbine at a rate of 1 kg/s and leaves at 0.2 bar. Neglect the KE and PE changes and estimate the power output of the turbine. Solve the problem using steam tables.

6.15 Rework Problem 6.14 using the Mollier diagram.

6.16 Steam at 3 bar and 240°C enters an adiabatic nozzle with negligible velocity and leaves at 1 bar. Assuming that the nozzle is isentropic, determine the discharge velocity of steam. Use steam tables to solve the problem.

6.17 Rework Problem 6.16 using the Mollier diagram.

6.18 A throttling calorimeter is used to measure the quality of wet steam flowing through a pipeline, where the pressure is 5 bar. The conditions of steam in the calorimeter are 1 bar and 110°C. Using steam tables, determine the quality of wet steam.

6.19 Rework Problem 6.18 using the Mollier diagram.

6.20 If the exit pressure in a throttling calorimeter is 1 bar and it is required that the steam leaving the calorimeter after throttling should be superheated by at least 5°C above the saturation temperature, estimate the minimum quality of wet steam which can be measured if the wet steam is at 30 bar. Make use of the Mollier diagram to solve the problem.

6.21 Rework Problem 6.20 using steam tables.

6.22 The test results for a small steam turbine show that steam enters the turbine at 50 bar and 400°C and leaves at 1 bar and 120°C. Using the Mollier diagram, determine the isentropic efficiency of the turbine.

6.23 Rework Problem 6.22 using steam tables.

6.24 A thermal power plant operates on a Rankine cycle. Superheated steam at 50 bar and 400°C enters the turbine where it expands to the condenser pressure of 10 kPa. Assuming that the turbine is isentropic, determine the thermal efficiency of the power plant and the steam flow rate through the turbine if the power output of the plant is 300 MW.

6.25 Rework Problem 6.24 if the isentropic efficiency of the turbine in 0.8.

6.26 A thermal power plant operates on a Rankine cycle. Superheated steam at 30 bar and 350°C enters a turbine, the isentropic efficiency of which is 0.8, and leaves at 10 kPa. The condensate leaving the condenser is pumped to the steam generator. The isentropic efficiency of the pump is 0.8. Determine the thermal efficiency of the power plant and the rate of steam generation if the power output of the plant is 200 MW.

7 Gas Power Cycles

7.1 Air Standard Cycles

We have discussed the Rankine cycle which uses a condensable vapor as a working medium. A large class of engines, however, employ a gas as the working medium and these cycles are called *gas power cycles*. Familiar examples of gas power cycles are the Otto cycle and the diesel cycle. The engines employing these cycles are sometimes called *Internal combustion (IC) engines*, because in these engines energy is added through the process of combustion of a fuel inside the cylinder itself, rather than from an outside source. The fuels used in these engines are gasoline (petrol), diesel oil and gasohol (90–10 mixture of gasoline and alcohol). In this class of engines, the gas is predominantly air with some products of combustion like CO_2, H_2O, etc. The composition of the working fluid does not remain constant throughout the cycle. In the beginning of the cycle, the working fluid is either air or fuel-air mixture and in subsequent processes it is a mixture of air and combustion products. In actual engines, the working fluid does not undergo cyclic changes, but the working fluid is exhausted at the end of every cycle and a fresh sample of air or fuel-air mixture is taken in at the beginning of the next cycle. However, the engine operates on a mechanical cycle.

For thermodynamic analysis of these engines, the following approximations are made. The cycles which are analyzed based on these approximations are called air standard cycles.

1. The engine is assumed to operate on a closed cycle with a fixed mass of working fluid.
2. The composition of the working fluid does not undergo any change during the cycle.
3. The addition of energy due to combustion of fuel is replaced by a convenient process in which energy is added from an external source.
4. The exhaust of the combustion products is replaced by a convenient energy rejection process in which energy is rejected to external surroundings.
5. The working fluid is assumed to be air and the air behaves like an ideal gas with constant specific heat.

7.2 Air Standard Otto Cycle

Nikolaus A Otto (1832–1891) developed a technically feasible four-stroke engine and hence it was named after him as the Otto engine. The Otto engine uses gasoline as fuel. The Otto cycle is shown in Fig. 7.1. It consists of the following processes:

0-1: The piston executes an outward stroke and a fresh mixture of fuel-air is drawn into the cylinder at constant pressure.

1-2: The piston executes an inward stroke and the fuel-air mixture is isentropically compressed. At the end of the compression stroke, the fuel-air mixture is at high temperature and pressure.

2-3: At state 2, the fuel-air mixture is ignited by introducing an electric spark. Since the gasoline-air mixture at sufficiently high temperature is highly combustible, the combustion is assumed to be instantaneous. There is not enough time for the piston to move backward during combustion and hence, the combustion process is replaced by energy addition at constant volume.

3-4: During the power stroke, the piston executes an outward stroke and the combustion products undergo isentropic expansion.
4-1: The exhaust valve opens and the combustion products leave the cylinder. The pressure in the cylinder reduces. However, the cylinder is still full of combustion products at low pressure.
1-0: The piston executes an inward stroke and pushes the combustion products out of the cylinder at constant pressure, while the exhaust valve is still open.

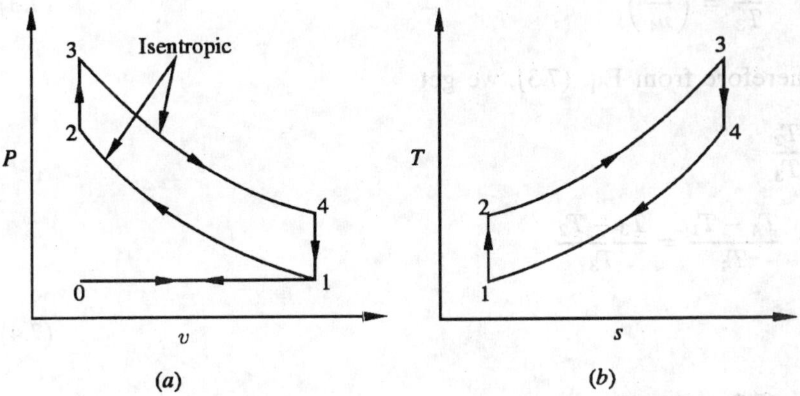

Fig. 7.1 *Air standard Otto cycle on (a) P—v diagram and(b) T—s diagram.*

In Fig. 7.1(a), the P–v diagram shows all the four strokes. The work done during the suction stroke 0-1 is exactly balanced by the work done during the exhaust stroke 1-0. The net work done because of these two strokes is equal to zero. Hence, for thermodynamic analysis, we consider the cycle 1-2-3-4 only as shown in Fig. 7.1(b). The thermal efficiency η of the cycle is given by

$$\eta = \frac{W_{\text{net}}}{Q_1} = \frac{Q_1 - Q_2}{Q_1} \tag{7.1}$$

where Q_1 and Q_2 denote the energy absorbed and rejected as heat, respectively. Energy is absorbed during the constant volume process 2-3 and energy is rejected during the constant volume process 4-1. We know that for a constant volume process $Q = \Delta U$. If m is the mass of the air which is undergoing the cyclic process, we get

$$\Delta U = mC_v \Delta T$$

Hence, $Q_1 = U_3 - U_2 = mC_v(T_3 - T_2)$

$$Q_2 = U_4 - U_1 = mC_v(T_4 - T_1)$$

Therefore,

$$\eta = \frac{mC_v(T_3 - T_2) - mC_v(T_4 - T_1)}{mC_v(T_3 - T_2)} = 1 - \frac{T_4 - T_1}{T_3 - T_2} \tag{7.2}$$

The processes 1-2 and 3-4 are isentropic. For an ideal gas undergoing an isentropic process, we know that

$$Tv^{\gamma-1} = \text{constant}$$

Hence,

$$\frac{T_1}{T_2} = \left(\frac{v_2}{v_1}\right)^{\gamma-1} \quad \text{and} \quad \frac{T_4}{T_3} = \left(\frac{v_3}{v_4}\right)^{\gamma-1} \tag{7.3}$$

But $v_1 = v_4$ and $v_2 = v_3$. Therefore from Eq. (7.3), we get

$$\frac{T_1}{T_2} = \frac{T_4}{T_3} \quad \text{or} \quad \frac{T_1}{T_4} = \frac{T_2}{T_3}$$

$$1 - \frac{T_1}{T_4} = 1 - \frac{T_2}{T_3} \quad \text{or} \quad \frac{T_4 - T_1}{T_4} = \frac{T_3 - T_2}{T_3}$$

$$\frac{T_4 - T_1}{T_3 - T_2} = \frac{T_4}{T_3} = \frac{T_1}{T_2} \tag{7.4}$$

Substituting Eq. (7.4) in Eq. (7.2), we get

$$\eta = 1 - \frac{T_1}{T_2} \tag{7.5}$$

From Eqs. (7.3) and (7.5), we get

$$\eta = 1 - \left(\frac{v_2}{v_1}\right)^{\gamma-1} = 1 - \left(\frac{1}{r_0}\right)^{\gamma-1} \tag{7.6}$$

where the compression ratio r_0, is defined as

$$r_0 = \frac{v_1}{v_2} \tag{7.7}$$

Since $\gamma > 1$, the efficiency of the Otto cycle increases with increasing compression ratio. However, in an actual engine the compression ratio cannot be increased indefinitely since higher compression ratios give higher temperatures of the gas after compression, and high temperature results in auto-ignition and uncontrolled combustion of the fuel. In the Otto cycle, the compression ratio is usually limited to 8.

Though the performance of an engine is expressed in terms of thermal efficiency, sometimes it is convenient to express it in terms of *Mean Effective Pressure* P_m, which is defined as

$$P_m = \frac{\text{Net work done}}{\text{Displacement volume}} = \frac{W}{v_1 - v_2} \tag{7.8}$$

$$W = P_m(v_1 - v_2) \tag{7.9}$$

EXAMPLE 7.1 In an Otto engine with a compression ratio of 8, the pressure and temperature of the gas at the beginning of the suction stroke are 1 bar and 300 K. During the combustion process, 1500 kJ of energy is added per kg of air. Calculate the temperature and pressure at the end of each process and the thermal efficiency of the engine.

SOLUTION Given: $T_1 = 300$ K; $P_1 = 1$ bar and $r_0 = 8$

Refer to Fig. 7.1 for the P–v and T–s diagrams of the Otto cycle.

Since process 1-2 is isentropic, we have

$$P_2 = P_1(v_1/v_2)^{\gamma} = 1(8)^{1.4} = 18.38 \text{ bar}$$

$$T_2 = T_1(v_1/v_2)^{\gamma-1} = 300(8)^{0.4} = 689.2 \text{ K}$$

2-3 is a constant volume process. Hence,

$$q_1 = u_3 - u_2 = C_v(T_3 - T_2)$$

$$1500 = 0.718(T_3 - 689.2) \quad \text{or} \quad T_3 = 2778.3 \text{ K}$$

$$P_3 = \frac{T_3}{T_2}P_2 = \frac{2778.3}{689.2} \times 18.38 = 74.09 \text{ bar}$$

3-4 is an isentropic process. Therefore,

$$T_4 = T_3\left(\frac{v_3}{v_4}\right)^{\gamma-1} = T_3\left(\frac{v_2}{v_1}\right)^{\gamma-1} = T_3\left(\frac{1}{r_0}\right)^{\gamma-1}$$

or $\quad T_4 = 2778.3\left(\frac{1}{8}\right)^{0.4} = 1209.3 \text{ K}$

$$P_4 = P_3\left(\frac{v_3}{v_4}\right)^{\gamma} = P_3\left(\frac{v_2}{v_1}\right)^{\gamma} = P_3\left(\frac{1}{r_0}\right)^{\gamma}$$

or $\quad P_4 = 74.09\left(\frac{1}{8}\right)^{1.4} = 4.03 \text{ bar}$

$$\text{Thermal efficiency, } \eta = 1 - \left(\frac{1}{r_0}\right)^{\gamma-1} = 1 - \left(\frac{1}{8}\right)^{0.4} = 0.565$$

EXAMPLE 7.2 Calculate the mean effective pressure for the Otto engine of Example 7.1.

SOLUTION Thermal efficiency, $\eta = 0.565$ (see Example 7.1). Work done, $W = Q_1\eta = 1500 \times 10^3 \times 0.565 = 847.5$ kJ

$$v_1 = \frac{RT_1}{P_1} = \frac{8.314 \times 10^3 \times 300}{28.97 \times 1 \times 10^5} = 0.861 \text{ m}^3\text{/kg}$$

$$v_2 = \frac{v_1}{r_0} = \frac{0.861}{8} = 0.1076 \text{ m}^3/\text{kg}$$

$$P_m = \frac{W}{v_1 - v_2} = \frac{847.5 \times 10^3}{0.861 - 0.1076} = 11.25 \text{ bar}$$

EXAMPLE 7.3 In an Otto engine, the conditions of air at the beginning of the suction stroke are 1 bar and 300 K. The maximum temperature and pressure in the cycle are 2727°C and 80 bar, respectively. Calculate the compression ratio and the thermal efficiency of the engine.

SOLUTION In the Otto cycle, the maximum temperature and pressure will be at state 3. Compression ratio,

$$r_0 = \frac{v_1}{v_2} = \frac{P_2 T_1}{T_2 P_1}$$

For the constant volume process 2-3, we have

$$\frac{P_2}{T_2} = \frac{P_3}{T_3}$$

Therefore,

$$r_0 = \frac{P_2 T_1}{T_2 P_1} = \frac{P_3 T_1}{T_3 P_1} = \frac{80 \times 300}{3000 \times 1} = 8$$

$$\text{Thermal efficiency, } \eta = 1 - \left(\frac{1}{r_0}\right)^{\gamma - 1} = 1 - \left(\frac{1}{8}\right)^{0.4} = 0.565$$

EXAMPLE 7.4 In an air standard Otto cycle, the net work done is 1000 kJ/kg air and the maximum temperature in the cycle is 3200 K. The temperature of the gas at the end of the isentropic compression is 800 K. Calculate the compression ratio of the cycle.

SOLUTION Given W = 1000 kJ/kg; T_2 = 800 K and T_3 = 3200 K.

$$q_1 = C_v (T_3 - T_2) = 0.718 (3200 - 800) = 1723.2 \text{ kJ/kg}$$

$$\eta = \frac{W}{q_1} = \frac{1000}{1723.2} = 0.58$$

$$\text{Thermal efficiency, } \eta = 1 - \left(\frac{1}{r_0}\right)^{\gamma - 1}$$

$$\text{or} \quad 0.58 = 1 - \left(\frac{1}{r_0}\right)^{0.4} \quad \text{or} \quad r_0 = 8.75$$

7.3 Air Standard Diesel Cycle

The diesel cycle was developed by Rudolf Diesel (1858–1913). This cycle is extensively used in present-day transport like in buses, trucks, cars and locomotives. High compression ratios can be used in the diesel cycle. In the diesel cycle only air is taken into the cylinder during the suction stroke. Then the air is compressed to a high pressure and temperature. The fuel (diesel oil) is then injected into the high temperature air for combustion. The combustion of the fuel occurs in the cylinder because of the high temperature of the air achieved by compression. Hence, the engine, which employs this cycle is quite often referred to as the *Compression–Ignition (CI)* engine. The diesel cycle shown in Fig. 7.2 consists of the following processes:

0-1: During the suction stroke, the piston executes an outward motion and fresh air is drawn into the cylinder.

1-2: During the compression stroke, the air is isentropically compressed to a high pressure (with a large compression ratio). Because of the large compression ratio, the temperature of the air at the end of the compression stroke will be very high.

2-3: The fuel injection starts at state 2. The temperature of the air at state 2 is so high that the combustion of the fuel occurs and the piston starts moving backwards. The combustion of the fuel (diesel oil) is not as instantaneous as the combustion of petrol and takes a finite time allowing the piston to move backwards. The energy addition by combustion is approximated as constant pressure energy addition.

3-4: During the power stroke or expansion stroke, the hot combustion products undergo isentropic expansion and the gas pressure in the cylinder reduces.

4-1: The exhaust valve opens at state 4, allowing the combustion products to leave the cylinder. This process is approximated as constant volume energy rejection to the surroundings.

1-0: The combustion products remaining in the cylinder are exhausted at constant pressure when the piston executes the inward motion.

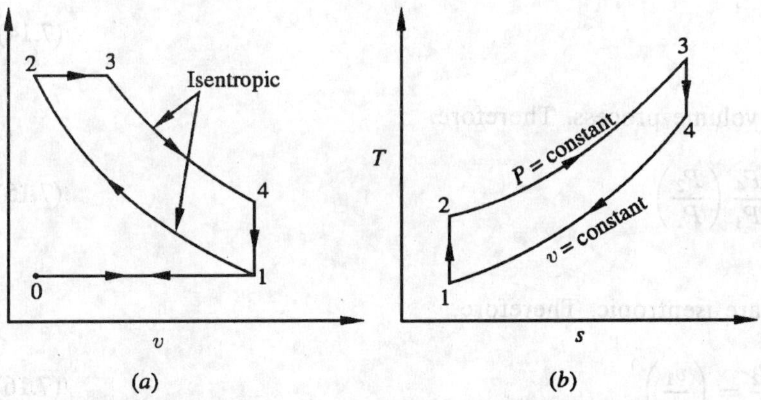

Fig. 7.2 Diesel cycle on (a) P–v diagram and (b) T–s diagram.

It can be seen from Fig. 7.2(a) that the net work done during the strokes 0-1 and 1-0 is zero. Therefore, we need to consider the cycle 1-2-3-4 as shown in Fig. 7.2(b), for thermodynamic analysis.

Now, let us evaluate the thermal efficiency of a diesel cycle. For this purpose, we define the cutoff ratio r_c, as

$$r_c = \frac{v_3}{v_2} \tag{7.10}$$

The energy addition takes place during the constant pressure process 2-3. For a constant pressure process, we know that $Q = \Delta H$. That is,

$$Q_1 = H_3 - H_2 = m(h_3 - h_2) = mC_p(T_3 - T_2) \tag{7.11}$$

where m is the mass of gas undergoing the cyclic change. The energy rejection takes place during the constant volume process. Hence,

$$Q_2 = U_4 - U_1 = m(m_4 - u_1) = mC_v(T_4 - T_1) \tag{7.12}$$

The thermal efficiency η is given by

$$\eta = \frac{Q_1 - Q_2}{Q_1} = \frac{mC_p(T_3 - T_2) - mC_v(T_4 - T_1)}{mC_p(T_3 - T_2)}$$

$$\eta = 1 - \frac{T_4 - T_1}{\gamma(T_3 - T_2)} = 1 - \frac{T_1\left(\dfrac{T_4}{T_1} - 1\right)}{T_2\gamma\left(\dfrac{T_3}{T_2} - 1\right)} \tag{7.13}$$

The process 1-2 is isentropic. Therefore,

$$\frac{T_1}{T_2} = \left(\frac{v_2}{v_1}\right)^{\gamma-1} = \left(\frac{1}{r_0}\right)^{\gamma-1} \tag{7.14}$$

The process 4-1 is a constant volume process. Therefore,

$$\frac{T_4}{T_1} = \frac{P_4}{P_1} = \frac{P_4}{P_3}\left(\frac{P_3}{P_1}\right) = \frac{P_4}{P_3}\left(\frac{P_2}{P_1}\right) \tag{7.15}$$

since $P_2 = P_3$.

The processes 1-2 and 3-4 are isentropic. Therefore,

$$\frac{P_4}{P_3} = \left(\frac{v_3}{v_4}\right)^{\gamma} \quad \text{and} \quad \frac{P_2}{P_1} = \left(\frac{v_1}{v_2}\right)^{\gamma} \tag{7.16}$$

From Eqs. (7.15) and (7.16), we get

$$\frac{T_4}{T_1} = \left(\frac{v_3}{v_4}\right)^{\gamma}\left(\frac{v_1}{v_2}\right)^{\gamma} = \left(\frac{v_3}{v_2}\right)^{\gamma} = r_c^{\gamma} \tag{7.17}$$

For the constant pressure process 2-3, we get

$$\frac{T_3}{T_2} = \frac{v_3}{v_2} = r_c \tag{7.18}$$

From Eqs. (7.13), (7.17) and (7.18), we get

$$\eta = 1 - \frac{1}{\gamma r_0^{\gamma-1}}\left[\frac{r_c^\gamma - 1}{r_c - 1}\right] \tag{7.19}$$

Equation (7.19) shows that the thermal efficiency of a diesel cycle can be increased by increasing the compression ratio r_0 and also by decreasing the cutoff ratio r_c. The compression ratios used in a diesel cycle vary from 14 to 17. It can be shown that the mean effective pressure of an air standard diesel cycle is given by

$$P_m = P_1 \frac{[\gamma r_0^\gamma (r_c - 1) - r_0(r_c^\gamma - 1)]}{(\gamma - 1)(r_0 - 1)} \tag{7.20}$$

EXAMPLE 7.5 In an air standard diesel cycle with a compression ratio of 16 and a cutoff ratio of 3, the conditions of air at the beginning of the suction stroke are 1 bar and 300 K. Calculate the thermal efficiency of the cycle and the energy added per kg air.

SOLUTION Given: $r_0 = 16$; $r_c = 3$; $P_1 = 1$ bar and $T_1 = 300$ K. The efficiency of the cycle is given by

$$\eta = 1 - \frac{1}{\gamma r_0^{\gamma-1}}\left[\frac{r_c^\gamma - 1}{r_c - 1}\right]$$

$$= 1 - \frac{1}{1.4(16)^{0.4}}\left[\frac{3^{1.4} - 1}{3 - 1}\right] = 0.5693$$

Since 1-2 is an isentropic process, we get

$$T_2 = T_1\left(\frac{v_1}{v_2}\right)^{\gamma-1} = T_1(r_0)^{\gamma-1} = 300(16)^{0.4} = 909.4 \text{ K}$$

The process 2-3 is a constant pressure process. Therefore,

$$\frac{T_3}{T_2} = \frac{v_3}{v_2} = r_c$$

$$T_3 = T_2 r_c = 909.4 \times 3 = 2728.2 \text{ K}$$

The energy added (q_1) per kg air is given by

$$q = C_p(T_3 - T_2) = 1.005(2728.2 - 909.4) = 1827.9 \text{ kJ/kg}$$

EXAMPLE 7.6 In an air standard diesel cycle, the minimum and maximum temperatures are 300 K and 2700 K, respectively. The minimum and maximum pressures are 1 bar and 40 bar, respectively. Calculate the compression ratio, cutoff ratio and the thermal efficiency of the cycle.

SOLUTION The P–v diagram for the diesel cycle is shown in Fig. 7.3. It can be observed from Fig. 7.3 that the minimum temperature and pressure of the cycle correspond to state 1, and the maximum temperature and pressure of the cycle correspond to state 3.

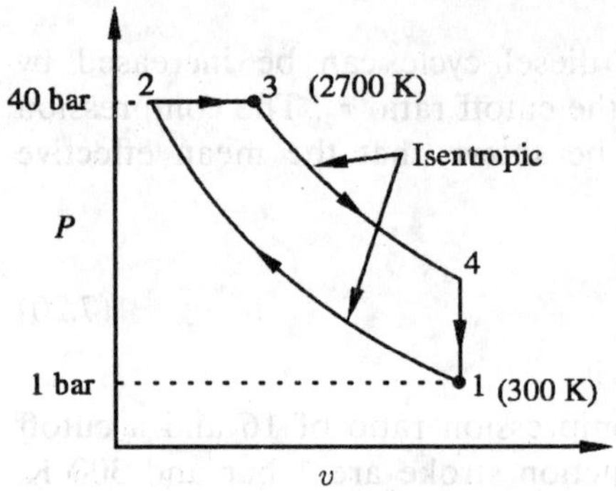

Fig. 7.3 P–v diagram for a diesel cycle.

$T_1 = 300$ K; $P_1 = 1$ bar; $T_3 = 2700$ K and $P_3 = 40$ bar; $P_2 = P_3 = 40$ bar

Since 1-2 is an isentropic process, we have

$$\frac{P_2}{P_1} = \left(\frac{v_1}{v_2}\right)^{\gamma} \quad \text{and} \quad \frac{T_2}{T_1} = \left(\frac{v_1}{v_2}\right)^{\gamma-1} = r_0^{\gamma-1}$$

$$\text{Therefore,} \quad \frac{40}{1} = r_0^{1.4} \quad \text{or} \quad r_0 = 13.94$$

$$T_2 = T_1 r_0^{\gamma-1} = 300(13.94)^{0.4} = 860.6 \text{ K}$$

The cutoff ratio, r_c is given by

$$r_c = \frac{v_3}{v_2}$$

For the constant pressure process 2-3, we have

$$\frac{v_3}{v_2} = r_c = \frac{T_3}{T_2} = \frac{2700}{860.6} = 3.137$$

The thermal efficiency η, is given by

$$\eta = 1 - \frac{1}{\gamma r_0^{\gamma-1}}\left[\frac{r_c^{\gamma}-1}{r_c-1}\right] = 1 - \frac{1}{1.4 \times (13.94)^{0.4}}\left[\frac{(3.137)^{1.4}-1}{3.137-1}\right] = 0.539$$

7.4 Comparison of Otto and Diesel Cycles

The Otto and diesel cycles can be compared on the basis of either the compression ratio or the maximum temperature and pressure of the cycle. Figure 7.4 shows the Otto and diesel cycle having the same compression ratio; both of them reject the same amount of energy Q_2 to the surroundings.

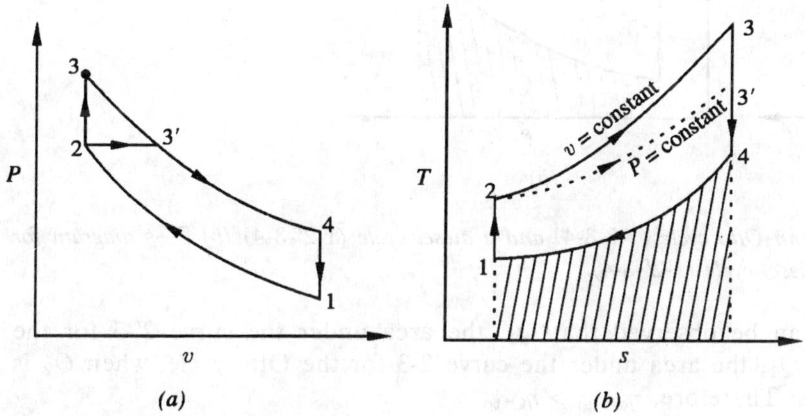

Fig. 7.4 (a) P–v diagram for an Otto cycle (1-2-3-4) and a diesel cycle(1-2-3′-4). (b) T–s diagram for an Otto cycle (1-2-3-4) and a diesel cycle (1-2-3′-4).

We know that

$$\int dQ = \int T ds$$

That is, the area under the curve on a T–s diagram represents the heat interaction. In Fig. 7.4, the Otto cycle is represented by 1-2-3-4 and the diesel cycle is represented by 1-2-3′-4. Both the cycles have the same compression ratio v_2/v_1 and reject the same quantity of energy Q_2 during the constant volume process 4-1 (hatched area under the curve 4-1 on the T–s diagram). The efficiency of the cycle is given by

$$\eta = 1 - \frac{Q_2}{Q_1}$$

For the same value of Q_2, the efficiency will be higher for higher values of Q_1. On the T–s diagram, the area under the curve 2-3 represents Q_1 for the Otto cycle and the area under the curve 2-3′ represents Q_1 for the diesel cycle. Since Q_1 for the Otto cycle is greater than that for the diesel cycle, we find that

$$\eta_{\text{Otto}} > \eta_{\text{diesel}}$$

The Otto and diesel cycles for the same maximum temperature and pressure (state 3) and for the same quantity of energy rejection Q_2 are shown in Fig. 7.5. The Otto cycle is represented by 1-2-3-4 and the diesel cycle is represented 1-2′-3-4.

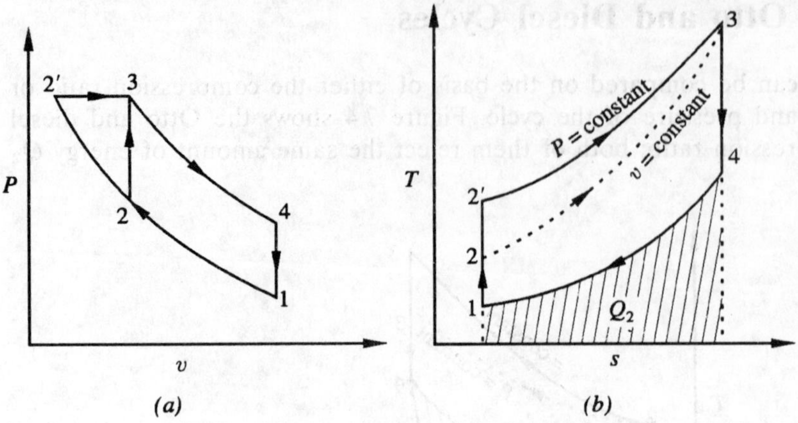

Fig. 7.5 (a) P–v diagram for an Otto cycle (1-2-3-4) and a diesel cycle (1-2′-3-4) (b) T–s diagram for an Otto cycle (1-2-3-4) and a diesel cycle (1-2′-3-4).

From the T–s diagram it can be observed that Q_1, the area under the curve 2′-3 for the diesel cycle, is greater than Q_1, the area under the curve 2-3 for the Otto cycle, when Q_2 is the same for both the cycles. Therefore, $\eta_{\text{diesel}} > \eta_{\text{Otto}}$

7.5 Air Standard Brayton Cycle

The Brayton cycle was developed by George B Brayton (1813–1892) and was first used with reciprocating engines. The Brayton cycle is ideally suited for gas turbines. The gas turbines operate on an open cycle. Atmospheric air is compressed to a high pressure in a compressor and then energy is added by the combustion of a fuel. Then the hot gases are expanded in a turbine to obtain work. Finally, the gas leaving the turbine is exhausted into the atmosphere. However, for thermodynamic analysis, we idealize the combustion process as a constant pressure energy addition from an external source, and the gas discharge from the turbine as replaced by an equivalent energy rejection at constant pressure, thus making the air undergo a cyclic process. The air standard Brayton cycle is shown in Fig. 7.6.

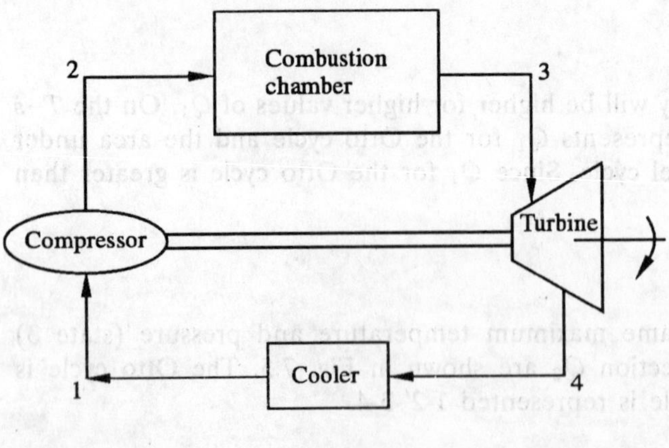

Fig. 7.6 Schematic representation of an air standard Brayton cycle.

The air standard Brayton cycle consists of the following processes:

1-2: Isentropic compression of air from pressure P_1 to P_2
2-3: Constant pressure energy addition at P_2 by combustion of a fuel
3-4: Isentropic expansion of hot gases from pressure P_2 to P_1 in a turbine.
4-1: Constant pressure energy rejection at P_1 in a cooler.

The $P–v$ and $T–s$ diagrams of the air standard Brayton cycle are shown in Fig. 7.7.

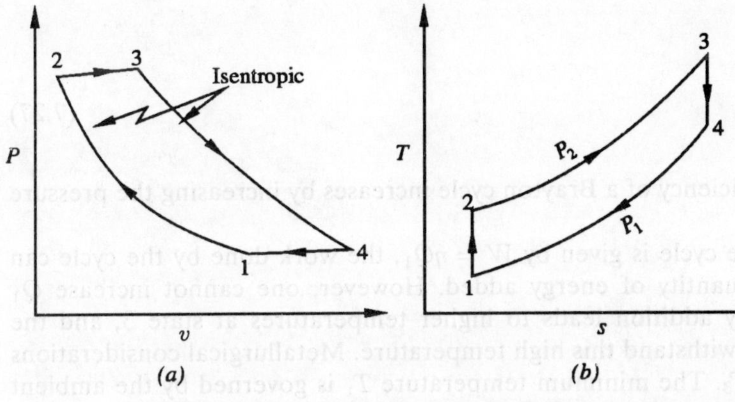

Fig. 7.7 (a) P–v diagram and (b) T–s diagram of the air standard Brayton cycle.

Let us now evaluate the thermal efficiency of the Brayton cycle.

Energy added, $Q_1 = mC_p(T_3 - T_2)$

Energy rejected, $Q_2 = mC_p(T_4 - T_1)$

$$\text{Thermal efficiency, } \eta = \frac{Q_1 - Q_2}{Q_1} = 1 - \frac{T_4 - T_1}{T_3 - T_2} \tag{7.21}$$

$$\eta = 1 - \frac{T_1[(T_4/T_1) - 1]}{T_2[(T_3/T_2) - 1]} \tag{7.22}$$

The pressure ratio of the Brayton cycle r_p, is defined as

$$r_p = P_2/P_1 \tag{7.23}$$

$$\text{Then} \quad \frac{P_3}{P_4} = \frac{P_2}{P_1} \tag{7.24}$$

The processes 1-2 and 3-4 are isentropic. We know that for an ideal gas undergoing isentropic process,

$$\frac{T_2}{T_1} = \left(\frac{P_2}{P_1}\right)^{(\gamma-1)/\gamma} \quad \text{and} \quad \frac{T_3}{T_4} = \left(\frac{P_3}{P_4}\right)^{(\gamma-1)/\gamma} \tag{7.25}$$

From Eqs. (7.24) and (7.25), we get

$$\frac{T_2}{T_1} = \frac{T_3}{T_4} \quad \text{or} \quad \frac{T_4}{T_1} = \frac{T_3}{T_2} \tag{7.26}$$

Substituting Eq. (7.26) in Eq. (7.22), we get

$$\eta = 1 - \frac{T_1}{T_2} = 1 - \left(\frac{P_1}{P_2}\right)^{(\gamma-1)/\gamma}$$

$$\eta = 1 - \left(\frac{1}{r_p}\right)^{(\gamma-1)/\gamma} \tag{7.27}$$

Equation (7.27) shows that the efficiency of a Brayton cycle increases by increasing the pressure ratio.

Since the work delivered by the cycle is given by $W = \eta Q_1$, the work done by the cycle can be increased by increasing the quantity of energy added. However, one cannot increase Q_1 indefinitely because higher energy addition leads to higher temperatures at state 3, and the turbine blades may not be able to withstand this high temperature. Metallurgical considerations limit the maximum temperature T_3. The minimum temperature T_1 is governed by the ambient atmosphere. Hence, for fixed values of T_1 and T_3, what is the optimum pressure ratio which yields the maximum work? Let us evaluate the optimum pressure ratio. The net work done by the cycle is given by

$$W_{\text{net}} = Q_1 - Q_2 = mC_p(T_3 - T_2) - mC_p(T_4 - T_1)$$

$$= mC_p\left[(T_3 - T_4) - (T_2 - T_1)\right] = mC_p\left[T_3\left(1 - \frac{T_4}{T_3}\right) - T_1\left(\frac{T_2}{T_1} - 1\right)\right] \tag{7.28}$$

Substituting Eq. (7.25) in Eq. (7.28), we get

$$W_{\text{net}} = mC_p\left[T_3\left(1 - \left(\frac{1}{r_p}\right)^{(\gamma-1)/\gamma}\right) - T_1\left((r_p)^{(\gamma-1)/\gamma} - 1\right)\right]$$

For optimum pressure ratio

$$\frac{dW_{\text{net}}}{dr_p} = mC_pT_3\left(\frac{\gamma-1}{\gamma}\right)(r_p)^{(1-2\gamma)/\gamma} - mC_pT_1\left(\frac{\gamma-1}{\gamma}\right)(r_p)^{-1/\gamma} = 0$$

$$\text{or} \quad T_3(r_p)^{(1-2\gamma)/\gamma} = T_1(r_p)^{-1/\gamma} \quad \text{or} \quad \frac{T_3}{T_1} = (r_p)^{2(\gamma-1)/\gamma}$$

$$\text{or} \quad r_p = \left(\frac{T_3}{T_1}\right)^{\gamma/2(\gamma-1)} \tag{7.29}$$

EXAMPLE 7.7 In an air standard Brayton cycle with a compression ratio of 8, the conditions of the air at the inlet of the compressor are 1 bar and 27°C. If the maximum allowed temperature in the cycle is 1000 K, calculate the thermal efficiency of the cycle and the net work done per kg air.

SOLUTION The thermal efficiency is given by

$$\eta = 1 - \left(\frac{1}{r_p}\right)^{(\gamma-1)/\gamma} = 1 - (1/8)^{0.4/1.4} = 0.448$$

We know that $T_2 = T_1\left(\frac{P_2}{P_1}\right)^{(\gamma-1)/\gamma} = T_1(r_p)^{(\gamma-1)/\gamma} = 300(8)^{0.4/1.4} = 543.4 \text{ K}$

Energy added, $Q_1 = mC_p(T_3 - T_2) = 1 \times 1.005(1000 - 543.4) = 458.88 \text{ kJ/kg}$

Net work done, $W = \eta Q_1 = 0.448 \times 458.88 = 205.6 \text{ kJ/kg}$

EXAMPLE 7.8 Calculate the optimum pressure ratio for the Brayton cycle of Example 7.7.

SOLUTION The optimum pressure ratio is given by

$$r_p = \left(\frac{T_3}{T_1}\right)^{\gamma/2(\gamma-1)} = \left(\frac{1000}{300}\right)^{1.4/(2\times0.4)} = 8.22$$

7.6 Indicator Diagram

An indicator diagram is a trace made by a pressure monitoring device which is attached to the cylinder of a reciprocating engine. The diagram shows the variation of gas pressure in the cylinder as a function of the volume of the gas in the cylinder. A typical indicator diagram appears as shown in Fig. 7.8.

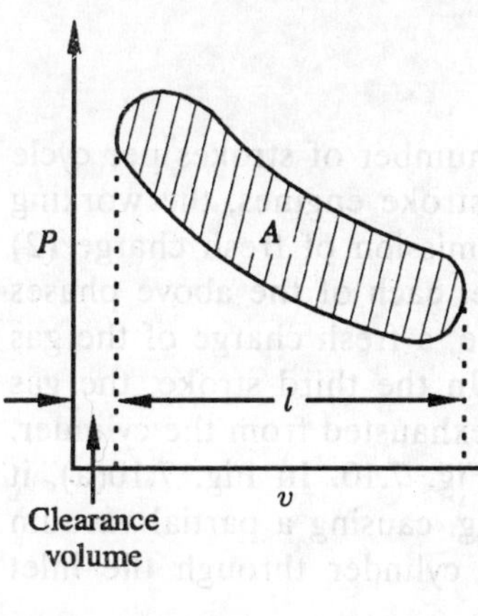

Fig. 7.8 Typical indicator diagram.

The area of the indicator diagram which is equal to the net work done per cycle is measured with a planimeter. If l denotes the length of the diagram and A denotes the area of the diagram, then the mean effective pressure P_m, is given by

$$P_m = \frac{AK}{l}$$

where K is the indicator spring constant. The length l of the diagram shows the stroke of the piston. The volume v_1 into which the gas is compressed during the compression stroke is called the *clearance volume* and the inside diameter of the cylinder is usually called the *bore*. Knowing the clearance volume, bore and stroke length, one can calculate v_2 and hence the compression ratio r_0. The bore, stroke and clearance volume are shown in Fig. 7.9.

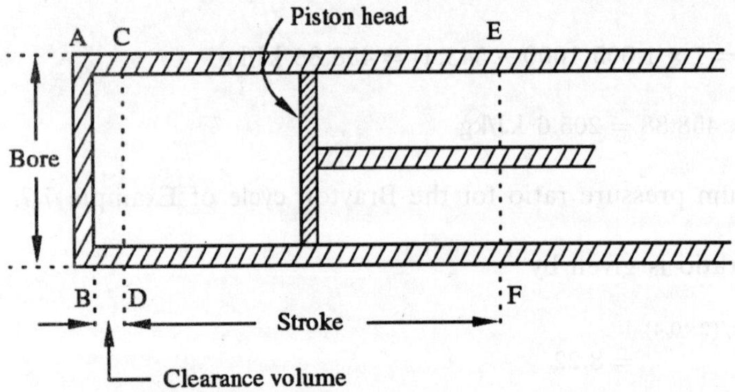

Fig. 7.9 Schematic of a piston-cylinder assembly showing bore, stroke and clearance volume.

During the compression stroke, the piston moves to the location CD and the gas is compressed into the volume represented by the region ABCD. The volume of the cylinder between the sections AB and CD is called the clearance volume. During the expansion stroke, the piston head moves up to the section EF. That is, the piston moves back and forth between sections CD and EF, and the length of the cylinder between these sections is called the stroke.

7.7 2-Stroke and 4-Stroke Engines

Internal combustion (IC) engines may be classified based on the number of strokes per cycle as 2-stroke and 4-stroke engines. In both the 4-stroke and the 2-stroke engines, the working cycle consists of four separate and distinct phases, namely (1) admission of fresh charge (2) compression (3) combustion and (4) exhaust. In a 4-stroke engine, each of the above phases is carried out in one complete stroke. That is, in the suction stroke, a fresh charge of the gas is admitted and in the second stroke, the charge is compressed. In the third stroke, the gas expands delivering work and finally in the fourth stroke, the gas is exhausted from the cylinder. The sequence of operations in a 4-stroke engine are shown in Fig. 7.10. In Fig. 7.10(a), it can be seen that during the suction stroke the piston is descending, causing a partial vacuum in the cylinder which induces the fresh charge to flow into the cylinder through the inlet valve (IV). The valve then closes and the piston returns to the original position, compressing the gas in the compression stroke as shown in Fig. 7.10(b). At the end of the compression stroke, the fuel-air mixture is ignited by the spark–ignition plug (SP) which increases the gas pressure. The increased gas pressure moves the piston down (expansion stroke) as shown in Fig. 7.10(c). Finally, the exhaust valve (EV) opens and the combustion products are exhausted during the exhaust stroke as shown in Fig. 7.10(d). These four strokes will be completed in two revolutions of the crankshaft.

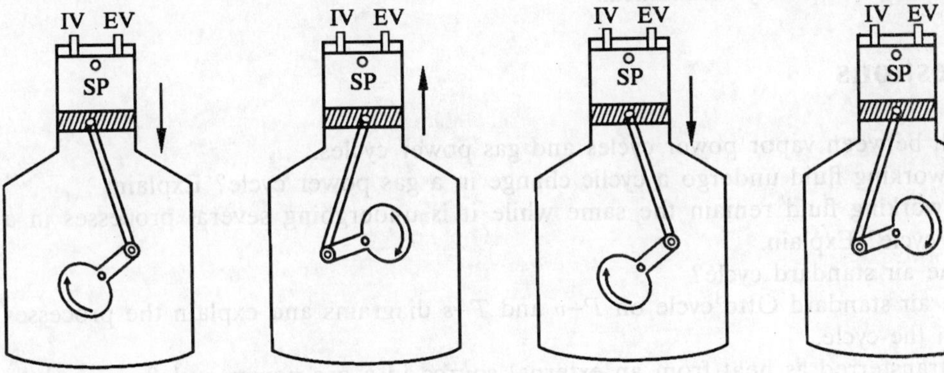

Fig. 7.10 Sequence of operations in a 4-stroke engine.

In a 2-stroke engine, all the four phases are carried out in a single revolution of the crankshaft. The working of a 2-stroke engine is illustrated in Fig. 7.11. In phase 1, the piston moves upwards while the exhaust port (EP) is uncovered by the piston. Then the gas exhausts from the cylinder and at the same time the transfer port (TP) opens and allows the fresh charge to enter the cylinder. In phase 2, as the piston rises further, the transfer port closes first and then the exhaust port closes and the charge is compressed. At this time a partial vacuum will

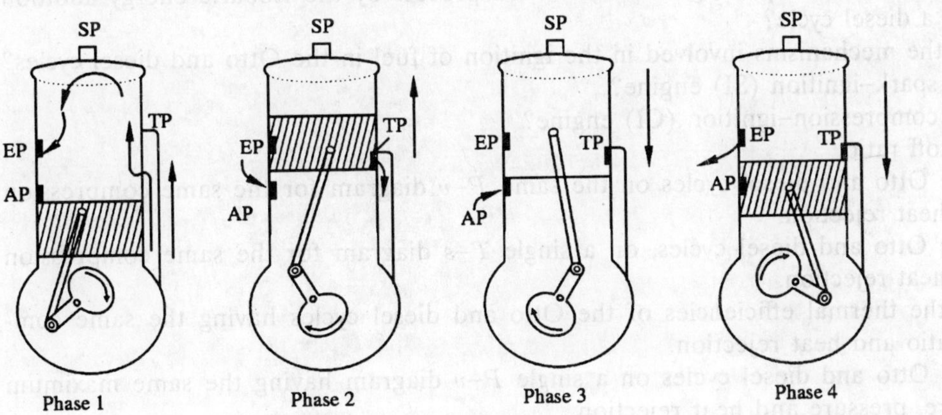

Fig. 7.11 Sequence of operations in a 2-stroke engine.

be created in the crankcase so that a charge is sucked into the space below the piston. In phase 3, the charge in the cylinder gets ignited by the spark plug (SP) and the combustion products force the piston to move downwards. During this process, the charge in the crankcase gets trapped by the closing of the admission port (AP) and gets compressed. In phase 4, the exhaust port (EP) opens and releases the combustion products from the cylinder, and the transfer port (TP) opens up allowing the charge in the crankcase to flow into the cylinder.

For thermodynamic analysis, there is no difference between a 2-stroke and a 4-stroke engine. The difference is purely mechanical.

Review Questions

1. Distinguish between vapor power cycles and gas power cycles.
2. Does the working fluid undergo a cyclic change in a gas power cycle? Explain.
3. Does the working fluid remain the same while it is undergoing several processes in a gas power cycle? Explain.
4. What is the air standard cycle?
5. Sketch the air standard Otto cycle on P–v and T–s diagrams and explain the processes involved in the cycle.
6. Is energy transferred as heat from an external source in a gas power cycle?
7. What is an internal combustion (IC) engine?
8. How is the energy added in an air standard Otto cycle?
9. What approximations are made in dealing with the energy rejection in an Otto cycle?
10. Define the compression ratio.
11. What is mean effective pressure?
12. How does the thermal efficiency of an Otto cycle vary with the compression ratio?
13. Sketch the air standard diesel cycle on P–v and T–s diagrams and explain the various processes undergone by the working fluid.
14. How does the energy addition in a diesel cycle differ from that in an Otto cycle?
15. What is the basis for replacing the combustion process by the isobaric energy addition process in a diesel cycle?
16. What are the mechanisms involved in the ignition of fuel in the Otto and diesel cycles?
17. What is a spark–ignition (SI) engine?
18. What is a compression–ignition (CI) engine?
19. Define cutoff ratio.
20. Sketch the Otto and diesel cycles on the same P–v diagram for the same compression ratio and heat rejection.
21. Sketch the Otto and diesel cycles, on a single T–s diagram for the same compression ratio and heat rejection.
22. Compare the thermal efficiencies of the Otto and diesel cycles having the same compression ratio and heat rejection.
23. Sketch the Otto and diesel cycles on a single P–v diagram having the same maximum temperature, pressure and heat rejection.
24. Compare the thermal efficiencies of the Otto and diesel cycles having the same heat rejection, maximum temperature and pressure in the cycle.
25. What is an indicator diagram?

26. What is the bore?
27. What is the stroke and clearance volume?
28. Explain (with the help of a neat sketch) the working principle of a 2-stroke engine.
29. Explain the working principle of a 4-stroke engine.
30. Sketch the air standard Brayton cycle on P–v and T–s diagrams and explain the working principle.
31. Is the efficiency of a Brayton cycle comparable with that of a Carnot cycle? Explain.
32. Define the pressure ratio of a Brayton cycle.
33. What is the dependence of thermal efficiency on the pressure ratio for a Brayton cycle?
34. Why is it not possible to operate a Brayton cycle with a large amount of energy addition?
35. Derive a relation to estimate the optimum pressure ratio for fixed values of minimum and maximum temperatures for a Brayton cycle.

Problems

7.1 In an air standard Otto cycle with a compression ratio of 8, the conditions of air at the start of the suction stroke are 1 bar and 300 K. During the combustion process 2000 kJ of energy is added per kg air. Determine (a) the temperature and pressure at the end of each process of the cycle. (b) the thermal efficiency of the cycle, and (c) the work done per kg air.

7.2 Determine the mean effective pressure of the Otto cycle of Problem 7.1.

7.3 In an air standard Otto cycle with a compression ratio of 8, the pressure and temperature at the end of the compression stroke are 15 bar and 700 K, respectively. The pressure rises to 40 bar after combustion. Calculate the thermal efficiency of the cycle and the net work done per kg air.

7.4 In an air standard Otto cycle, the maximum temperature in the cycle is 3200 K and the temperature at the end of the compression stroke is 800 K. If the net work done is 900 kJ/kg air, estimate the compression ratio.

7.5 In an air standard Otto cycle, the maximum temperature and pressure of the cycle are 2800°C and 70 bar, while air at 1 bar and 27°C is drawn into the cylinder at the beginning of the compression stroke. Calculate the thermal efficiency of the cycle.

7.6 Estimate the mean effective pressure for the Otto cycle of Problem 7.5.

7.7 In an air standard Otto cycle, the minimum and maximum allowed temperatures are T_1 and T_3, respectively. Determine the compression ratio which gives the maximum net work.

7.8 In an air standard diesel cycle with a compression ratio of 16, the conditions of air in the cylinder at the beginning of the compression stroke are 1 bar and 27°C. If the maximum temperature attained after energy addition is 1727°C, calculate the thermal efficiency of the cycle.

7.9 For the diesel cycle of Problem 7.8, estimate the mean effective pressure.

7.10 In an air standard diesel cycle, the conditions of the air at the start of the compression stroke are 1 bar and 27°C. The pressure rises to 40 bar after compression, and through combustion, 1200 kJ of energy is added per kg air. Calculate the compression ratio, the cutoff ratio and the thermal efficiency of the cycle.

7.11 Show that the mean effective pressure P_m of an air standard diesel cycle is given by

$$P_m = P_1 \frac{\gamma r_0^\gamma (r_c - 1) - r_0 (r_c^\gamma - 1)}{(\gamma - 1)(r_0 - 1)}$$

7.12 In a diesel cycle, the temperatures before and after the compression stroke are 27°C and 627°C, respectively. The energy addition during the combustion process is 900 kJ/kg air. Determine the compression ratio, the cutoff ratio and the maximum temperature in the cycle.

7.13 In an air standard Brayton cycle with a pressure ratio of 6, the conditions of the air at the inlet of the compressor are 1 bar and 27°C. During the combustion process, 450 kJ of energy is added per kg air. Estimate the maximum temperature of the cycle and the work done per kg air.

7.14 In an air standard Brayton cycle, if the temperature at the beginning and end of the compression are 27°C and 327°C, estimate the pressure ratio and thermal efficiency of the cycle.

7.15 A Brayton cycle is to be operated with minimum and maximum temperatures of 27°C and 727°C. What should the pressure ratio be so that the net work done per cycle is maximum?

7.16 The maximum allowed temperature in a Brayton cycle is 1200 K. If the air at the start of the compression is at 300 K and 1 bar and the pressure ratio of the cycle is 5, calculate the work done by the turbine and the energy added as heat per kg air.

8 VAPOR COMPRESSION REFRIGERATION

Refrigeration is the production and maintenance of a temperature lower than the ambient atmospheric temperature. The device used for transferring energy from a low temperature body to a high temperature body is called a refrigerator. The refrigerator is essentially a heat engine operating in the reverse order. The coefficient of performance (COP) of a refrigerator is defined as

$$(\text{COP})_{\text{R}} = \frac{\text{Energy absorbed from the low temperature body}}{\text{Work done}}$$

$$\frac{Q_L}{W} = \frac{Q_L}{Q_H - Q_L} \tag{8.1}$$

A schematic of a refrigerator is shown in Fig. 8.1.

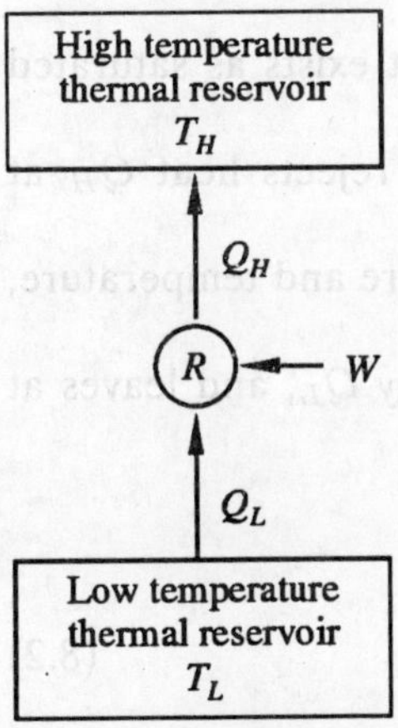

Fig. 8.1 . Schematic representation of a refrigerator.

The refrigeration cycles are classified as vapor compression refrigeration cycles and gas refrigeration cycles, depending on whether the working medium is a condensable vapor or gas. In this chapter the vapor compression refrigeration cycle is discussed.

8.1 Carnot Vapor Compression Refrigeration Cycle

The Second Law of Thermodynamics tells us that the coefficient of performance (COP) of a refrigerator working between two thermal reservoirs with fixed temperatures cannot be infinity. It also tells us that no refrigerator can have a COP greater than the COP of a Carnot refrigerator, when both are operated between the same set of thermal reservoirs. Hence a Carnot refrigerator is taken as a standard of perfection against which the performance of any practical refrigerator can be compared.

Consider a Carnot refrigerator which employs a vapor as a working medium. A schematic of a Carnot vapor compression refrigerator is shown in Fig. 8.2.

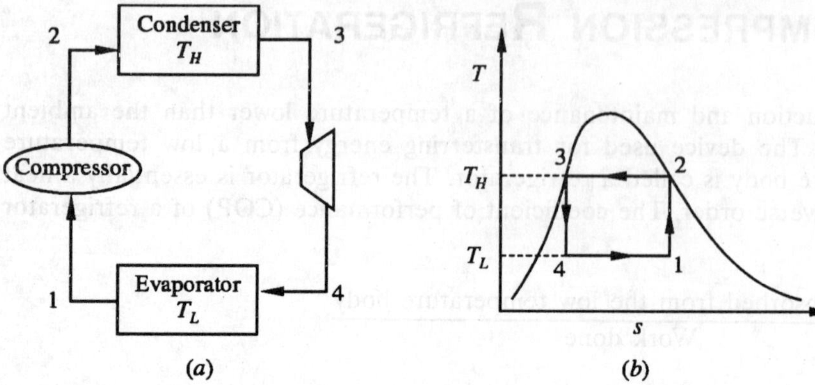

Fig. 8.2 (a) Schematic representation and (b) T–s diagram of a Carnot vapor compression refrigeration cycle.

The refrigeration cycle consists of the following processes:

1-2: Wet vapor at state 1 is isentropically compressed to state 2, where it exists as saturated vapor.
2-3: Saturated refrigerant vapor at state 2 enters a condenser where it rejects heat Q_H at temperature T_H.
3-4: The saturated liquid at state 3 is isentropically expanded to low pressure and temperature, and leaves at T_L.
4-1: The refrigerant at state 4 enters the evaporator at T_L, absorbs energy Q_L, and leaves at state 1.

The COP of the refrigerator is given by

$$(\text{COP})_\text{R} = \frac{Q_L}{W} = \frac{Q_L}{Q_H - Q_L} = \frac{T_L}{T_H - T_L} \tag{8.2}$$

Equation (8.2) shows that the COP is higher if the temperature difference between the reservoirs is small.

8.2 Practical Vapor Compression Refrigeration Cycle

The Carnot vapor compression refrigeration cycle is quite desirable. However, it is not practical. The isentropic expansion of saturated liquid through a turbine is not practical, because the turbine or expander has to handle a two-phase mixture of low quality. Application of the First Law of Thermodynamics to the control volume's expander and compressor gives

$$(-W_s/\dot{m})_{\text{Expander}} = h_4 - h_3 = \int_3^4 v\,dp \tag{8.3}$$

$$(-W_s/\dot{m})_{\text{Compressor}} = h_2 - h_1 = \int_1^2 v\,dp \tag{8.4}$$

Both the expander and compressor operate between the same pressure limits. However, the specific volume range in which the expander works is very much lower than the specific volume range in which the compressor operates. This can be observed from the T–s diagram. The process 3-4 occurs while the refrigerant is close to the saturated liquid curve, while the process 1-2 occurs while the refrigerant is close to the saturated vapor curve. Hence, the work done by the expander will be negligible compared to the work done on the compressor. In a practical refrigeration cycle, the expander is replaced by an inexpensive throttle valve. A schematic representation of the practical refrigeration cycle is shown in Fig. 8.3.

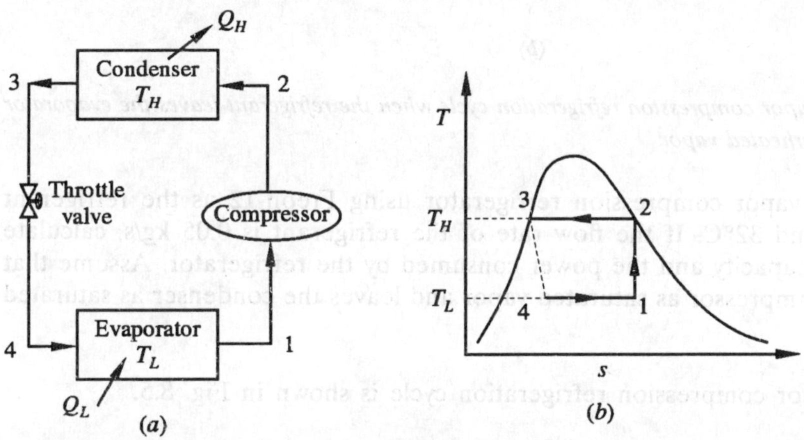

Fig. 8.3 (a) Schematic of a practical vapor compression refrigeration cycle. (b) T–s diagram of a vapor compression refrigeration cycle.

In a practical vapor compression refrigeration system, the saturated liquid at state 3 leaving the condenser flows through a throttle valve. During throttling, the liquid partially vaporizes and its temperature reduces to T_L. The refrigerant at state 4 leaving the throttle valve enters the evaporator and absorbs energy Q_L from the body at temperature T_L. Application of the First Law of Thermodynamics to the control volume compressor, condenser, throttle valve and evaporator gives

$$(W_s)_{\text{compressor}} = h_2 - h_1; \quad Q_H = h_2 - h_3$$

$$h_3 = h_4 \quad \text{and} \quad Q_L = h_1 - h_4$$

The COP of the refrigerator is then given by

$$(\text{COP})_{\text{R}} = \frac{Q_L}{W} = \frac{h_1 - h_4}{h_2 - h_1} \tag{8.5}$$

In the ideal refrigeration cycle, the refrigerant leaves the evaporator as wet vapor. In some cases, it is possible that the refrigerant leaving the evaporator is either saturated vapor or superheated vapor. Then the refrigeration cycle appears as shown in Fig. 8.4.

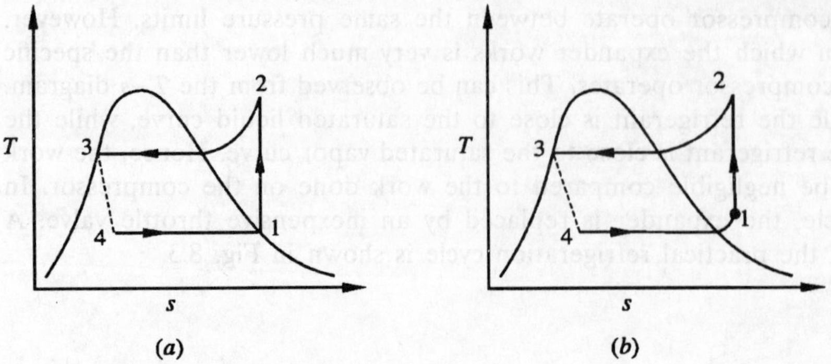

Fig. 8.4 T–s diagram for a vapor compression refrigeration cycle when the refrigerant leaves the evaporator as (a) saturated vapor. (b) superheated vapor.

EXAMPLE 8.1 An ideal vapor compression refrigerator using Freon-12 as the refrigerant operates between −22°C and 32°C. If the flow rate of the refrigerant is 0.05 kg/s, calculate the COP, the refrigeration capacity and the power consumed by the refrigerator. Assume that the refrigerant leaves the compressor as saturated vapor and leaves the condenser as saturated liquid.

SOLUTION The ideal vapor compression refrigeration cycle is shown in Fig. 8.5.

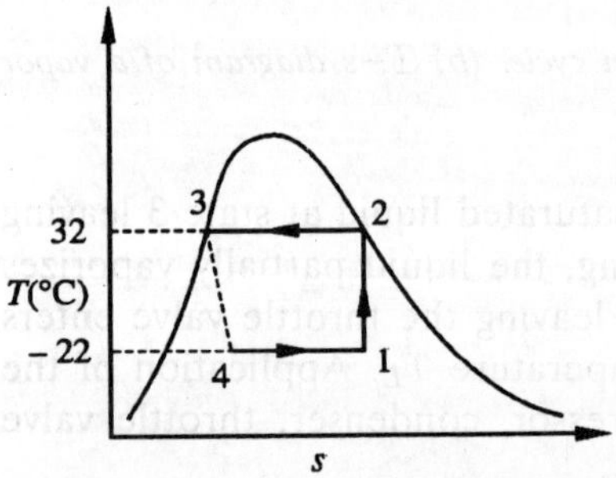

Fig. 8.5 Sketch for Example 8.1.

From the saturated refrigerant Freon-12 (temperature) table, we read the following values:

At $T = 32°C$:

$$h_f = h_3 = 67.0939 \text{ kJ/kg}; \quad h_g = h_2 = 200.3571 \text{ kJ/kg} \quad \text{and} \quad s_g = s_2 = 0.6855 \text{ kJ/kg K}$$

At $T = -22°C$:

$$h_f = 16.136 \text{ kJ/kg} \quad \text{and} \quad h_{fg} = 161.8590 \text{ kJ/kg}$$

$$s_f = 0.0665 \text{ kJ/kg K} \quad \text{and} \quad s_g = 0.7109 \text{ kJ/kg K}$$

1-2 is an isentropic process. Therefore,

$$s_2 = s_1 = s_f(1 - X_1) + s_g X_1$$

$$0.6855 = 0.0665(1 - X_1) + 0.7109X_1 \quad \text{or} \quad X_1 = 0.958$$

$$h_1 = h_f + X_1 h_{fg} = 16.1366 + 0.958 \times 161.859 \quad \text{or} \quad h_1 = 170.94 \text{ kJ/kg}$$

3-4 is an isenthalpic process. Therefore, $h_3 = h_4$

Then $(\text{COP})_R = \dfrac{h_1 - h_4}{h_2 - h_1} = \dfrac{170.94 - 67.0939}{200.3571 - 170.94} = 3.505$

Refrigeration capacity, $\dot{Q}_L = \dot{m}(h_1 - h_4) = 0.05(170.94 - 67.0939) = 5.19 \text{ kW}$

Power consumption, $\dot{W}_s = \dot{m}(h_2 - h_1) = 0.05(200.571 - 170.94) = 1.48 \text{ kW}$

EXAMPLE 8.2 In an ideal vapor compression refrigerator with Freon-12 as the refrigerant, the refrigerant leaves the compressor as saturated vapor at a pressure of 8 bar. The refrigerant pressure in the evaporator is 1.2 bar. Estimate the circulation rate of the refrigerant and the COP if the refrigeration capacity of the unit is 50 kW. Assume that the refrigerant leaves the condenser as saturated liquid.

SOLUTION Read the following values from the saturated refrigerant Freon-12 (pressure) table.

At P = 1.2 bar:

$$T = -25.7°\text{C}; \quad h_f = 12.7702 \text{ kJ/kg}; \quad h_{fg} = 163.5288 \text{ kJ/kg}$$

$$s_f = 0.0530 \text{ kJ/kg K} \quad \text{and} \quad s_g = 0.7139 \text{ kJ/kg K}$$

At P = 8 bar:

$$T = 32.8°\text{C}; \quad h_f = 67.8449 \text{ kJ/kg}; \quad h_g = 200.6255 \text{ kJ/kg} \quad \text{and} \quad s_g = 0.6848 \text{ kJ/kg K}$$

Refer to Fig. 8.5 for the T–s diagram of the refrigeration cycle.

$$h_2 = 200.6255 \text{ kJ/kg}; \quad s_2 = 0.6848 \text{ kJ/kg K}; \quad h_3 = h_4 = 67.8449 \text{ kJ/kg}$$

For an isentropic compressor,

$$s_2 = s_1 = s_f(1 - X_1) + s_g X_1$$

or $\quad 0.6848 = 0.0530(1 - X_1) + 0.7139X_1 \quad \text{or} \quad X_1 = 0.956$

$$h_1 = h_f + X_1 h_{fg} = 12.7702 + 0.956 \times 163.5288 = 169.1 \text{ kJ/kg}$$

$$\dot{q}_L = h_1 - h_4 = 169.1 - 67.8449 = 101.25 \text{ kJ/kg}$$

Refrigerant circulation rate,

$$\frac{\dot{Q}_L}{\dot{q}_L} = \frac{50}{101.25} = 0.494 \text{ kg/s}$$

$$(\text{COP})_R = \frac{h_1 - h_4}{h_2 - h_1} = \frac{169.1 - 101.25}{200.6255 - 169.1} = 2.15$$

Quite often the capacity of a refrigeration unit is expressed in terms of tons of refrigeration. If a refrigerating device can convert 1 ton (2000 pounds = 907.18 kg) of water at 0°C into ice at 0°C in 24 hours, its capacity is called one ton of refrigeration. Therefore, one ton of refrigeration

$$= \frac{907.18 \times 334.9}{24 \times 60} = 211 \text{ kJ/min}$$

(The latent heat of fusion of water at 0°C = 334.9 kJ/kg)

8.3 Refrigerants

In the vapor compression refrigeration cycles, the working fluid (or refrigerant) undergoes a phase change during a cycle. The choice of the substance to be used as a refrigerant is governed by the dependence of saturation pressure on the temperature of the substance under consideration. Refrigerators are used to maintain low temperatures of the order of – 20°C to – 30°C and hence, the refrigerant should undergo phase change transition at these low temperatures. Moreover, at these low temperatures it should have a saturation pressure of 1 bar or more. If the refrigerant is at a pressure lower than the atmospheric pressure in the evaporator, air may leak into the refrigerant. The refrigerant leaving the evaporator is compressed, and then it should reject energy to the surrounding atmosphere to undergo a phase change in the condenser. In summer, the atmospheric temperature may be as high as 45°C. Hence, the refrigerant should have a high critical pressure so that it can coexist in both the liquid and the vapor phases even at high pressures. Moreover, the refrigerant should be non-toxic and non-flammable because of the possibility of leakage into the room where the refrigerator is working. The refrigerant undergoes several cycles while in use and hence, it should be chemically stable for a long period of time, and at the same time it should not be too expensive.

Review Questions

1. What is refrigeration?
2. Define the coefficient of performance (COP) of a refrigerator.
3. How are refrigeration cycles classified?
4. Describe a Carnot vapor compression refrigeration cycle.
5. What are the practical difficulties associated with the operation of a Carnot vapor compression refrigeration cycle?

6. What modifications are made in the Carnot vapor compression refrigeration cycle while devising a practical refrigeration cycle?
7. Sketch the P–v and T–s diagrams of an ideal vapor compression refrigeration cycle.
8. Is there any loss in refrigeration capacity if the isentropic expansion of the refrigerant is replaced by the throttling process? Explain.
9. What is a ton of refrigeration?
10. What are the important factors which should be taken into account while selecting a refrigerant?

Problems

8.1 The COP of a Carnot refrigerator can be increased either by increasing the evaporator temperature while keeping the condenser temperature constant, or by decreasing the condenser temperature while the evaporator is held at constant temperature. Which of these alternatives is more effective?

8.2 Two identical bodies of mass m and specific heat C are available at the same temperature T_1. It is desired to operate a refrigerator between these two bodies to cool one of them to temperature T_2. Determine the minimum amount of work to be done.

8.3 An ideal vapor compression refrigerator using Freon-12 as the refrigerant is used to maintain certain drugs at – 22°C while the ambient temperature is 38°C. If the power consumption of the refrigerator is 1.5 kW, determine the capacity of the refrigerator. Assume that the refrigerant is in the saturated vapor state at the exit of the compressor and in the saturated liquid state at the exit of the condenser.

8.4 A 2-ton capacity ideal vapor compression refrigerator with Freon-12 as the refrigerant is operated to maintain a low temperature of – 30°C while the surrounding atmospheric air is at 38°C. Calculate the COP and power consumption of the refrigerator.

8.5 In an ideal vapor compression refrigeration cycle, Freon-12 is used as a refrigerant. The refrigerant pressure in the evaporator is 1.2 bar and in the condenser it is 10 bar. The circulation rate of the refrigerant is 0.5 kg/s. The condition of the refrigerant at the exit of the compressor is saturated vapor and at the exit of the condenser it is saturated liquid. Calculate the COP, the refrigeration capacity in tons of refrigeration and the power consumption.

9 PSYCHROMETRY

Quite often an engineer deals with mixtures of condensable vapor and gases. The combustion products in the cylinder of an IC engine contain water vapor in addition to CO_2, CO and other products. In chemical process industries, many of the gas streams contain condensable vapors. In the field of air-conditioning, an engineer deals with mixtures of air and water vapor. Therefore, we shall study the thermodynamics of mixtures of condensable vapor and gases (in particular air–water vapor mixtures) in this chapter.

9.1 Saturated Air–Water Vapor Mixture

At low pressures (below the atmospheric pressure), the condensable vapor may be approximated as an ideal gas. Suppose dry air is brought into contact with a pool of liquid water and sufficient time is allowed for the air to interact with water, part of the water evaporates and the water vapor diffuses into the air till the liquid water is in a state of equilibrium with the air–water vapor mixture. At equilibrium, the partial pressure of water vapor p_w in the air–water vapor mixture will be equal to the saturation pressure (or vapor pressure) P^s of water at the given temperature. However long the two phases—liquid water and air–water vapor mixture—are kept in contact, the partial pressure of water vapor in the air–water vapor mixture does not undergo any further change. Then the air–water vapor mixture is said to be saturated. In other words, for the saturated air–water vapor mixture,

$$p_w = P^s \tag{9.1}$$

9.2 Dew Point

Consider a mixture of air and water vapor which is unsaturated. That is, in the mixture the partial pressure of water vapor is less than the saturation pressure of water at the mixture temperature. If this mixture is cooled at constant pressure, the temperature of air as well as of water vapor keeps decreasing. The saturation pressure of water decreases as the temperature of the mixture decreases. Initially the partial pressure of water vapor in the mixture remains unchanged, but with continued cooling, a point will be reached where the saturation pressure of water becomes equal to the partial pressure of water vapor in the mixture. That is, at a particular temperature $p_w = P^s$. Dew forms at this temperature. The temperature at which the vapor starts condensing is called the dew point.

EXAMPLE 9.1 The partial pressure of water vapor in an air–water vapor mixture at 1 bar and 40°C is 0.025 bar. Calculate the dew point of this mixture.

SOLUTION From steam tables we find the saturation pressure of water at 40°C = 0.073 75 bar.

Given partial pressure of water vapor = p_w = 0.025 bar.

At dew point $p_w = P^s = 0.025$ bar.

From steam tables, we find $T = 20°C: P^s = 0.023\ 37$ bar

$T = 25°C: P^s = 0.031\ 66$ bar

By interpolation, we get $P^s = 0.025$ bar at $T = 20.98°C$

Hence, the dew point of the mixture = 20.98°C.

That is, the condensation of water vapor occurs if the mixture is cooled to 20.98°C.

9.3 Specific Humidity and Relative Humidity

The composition of air–water vapor mixtures can be expressed in terms of mass fractions or mole fractions of air and water vapor. However, for air–water vapor mixtures, it is customary to specify the composition in terms of specific humidity or relative humidity. The specific humidity or humidity ratio (ω) is defined as the ratio of the mass of water vapor (m_w) to the mass of dry air (m_a) in the mixture. That is,

$$\text{Specific humidity } \omega = \frac{m_w}{m_a}$$

At low pressure, the water vapor may be assumed to behave like an ideal gas. Then

$$\omega = \frac{m_w}{m_a} = \frac{(p_w V/RT)M_w}{(p_a V/RT)M_a} = \frac{p_w \times 18.02}{p_a \times 28.97} = 0.622\frac{p_w}{p_a} \tag{9.2}$$

$$\omega = \frac{0.622 p_w}{P - p_w} \tag{9.3}$$

where p_a, p_w = partial pressure of dry air and water vapor in the mixture, respectively; M_a, M_w = molar mass of dry air and water vapor, respectively; P = total pressure.

The relative humidity (ϕ) is defined as the ratio of the partial pressure of the water vapor in the mixture to the saturation pressure of water at the mixture temperature. That is,

$$\phi = \frac{p_w}{P^s} \tag{9.4}$$

The relative humidity and specific humidity are related to each other as

$$\omega = 0.622\frac{p_w}{p_a} = 0.622\frac{p_w}{P^s}\left(\frac{P^s}{p_a}\right) = 0.622\phi\frac{P^s}{p_a}$$

$$\phi = \frac{\omega p_a}{0.622 P^s} \tag{9.5}$$

EXAMPLE 9.2 Calculate the specific humidity and relative humidity of the air–water vapor mixture of Example 9.1.

SOLUTION Given $P = 1$ bar; $p_w = 0.025$ bar and $T = 40°C$. At 40°C, $P^s = 0.073\ 75$ bar.

$$\text{Therefore,} \quad \omega = \frac{0.622\, p_w}{P - P_w} = \frac{0.622 \times 0.025}{1 - 0.025} = 0.0159 \frac{\text{kg water vapor}}{\text{kg dry air}}$$

$$\phi = \frac{p_w}{P^s} = \frac{0.025}{0.073\ 75} = 0.339$$

or relative humidity = 33.9%.

EXAMPLE 9.3 Calculate the dew point and specific humidity for an air–water vapor mixture of 20% relative humidity at 1 bar and 40°C.

SOLUTION At 40°C, $P^s = 0.073\ 75$ bar

$$\phi = \frac{p_w}{P^s} \quad \text{or} \quad 0.2 = \frac{p_w}{0.073\ 75} \quad \text{or} \quad p_w = 0.014\ 75 \text{ bar.}$$

$$\omega = \frac{0.662 p_w}{P - p_w} = \frac{0.662 \times 0.014\ 75}{1 - 0.014\ 75} = 9.31 \times 10^{-3} \frac{\text{kg water vapor}}{\text{kg dry air}}$$

At dew point, $p_w = 0.014\ 75$ bar.

From steam tables, we find $T = 12.6°C$.

9.4 Adiabatic Saturation

The specific humidity of an air–water vapor mixture can be measured with the help of a device called an adiabatic saturator. Consider the device shown in Fig. 9.1 into which unsaturated air–water vapor mixture flows steadily and leaves as saturated air.

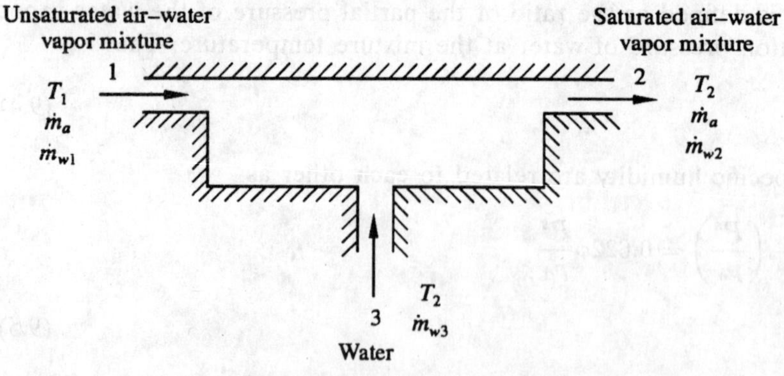

Fig. 9.1 Schematic representation of an adiabatic saturator

Let $\dot{m}_a$ = mass flow rate of dry air; $\dot{m}_{w1}, \dot{m}_{w2}$ = mass flow rate of water vapor along with dry air at the inlet (1) and at the exit (2) of the device; $\dot{m}_{w3}$ = mass flow rate of liquid water at the inlet (3) into the device.

The device is perfectly insulated such that there is no energy exchange as heat with the surroundings. Let us assume that liquid water at temperature T_2, which is the same as the temperature of the saturated air leaving the device, enters the device in a steady-state. The mass balance for water gives

$$\dot{m}_{w3} = \dot{m}_{w2} - \dot{m}_{w1}$$

$$\frac{\dot{m}_{w3}}{\dot{m}_a} = \frac{\dot{m}_{w2}}{\dot{m}_a} - \frac{\dot{m}_{w1}}{\dot{m}_a} = \omega_2 - \omega_1 \tag{9.6}$$

Energy balance gives

$$\dot{m}_a C_{pa} T_1 + \dot{m}_{w1} h_{g1} + \dot{m}_{w3} h_{f2} = \dot{m}_a C_{pa} T_2 + \dot{m}_{w2} h_{g2}$$

$$C_{pa} T_1 + \frac{\dot{m}_{w1}}{\dot{m}_a} h_{g1} + \frac{(\dot{m}_{w2} - \dot{m}_{w1})}{\dot{m}_a} h_{f2} = C_{pa} T_2 + \frac{\dot{m}_{w2}}{\dot{m}_a} h_{g2}$$

$$C_{pa} T_1 + \omega_1 h_{g1} + (\omega_2 - \omega_1) h_{f2} = C_{pa} T_2 + \omega_2 h_{g2} \tag{9.7}$$

where C_{pa} = specific heat of dry air; h_g = specific enthalpy of water vapor; h_{f2} = specific enthalpy of liquid water at T_2.

In the air–water vapor mixture, the water vapor exists in the superheated state, because it is at a temperature which is above the saturation temperature corresponding to the partial pressure or above the dew point. Equation (9.7) can be rearranged as

$$\omega_1 = \frac{C_{pa}(T_2 - T_1) + \omega_2 (h_{g2} - h_{f2})}{(h_{g1} - h_{f2})} = \frac{C_{pa}(T_2 - T_1) + \omega_2 h_{fg2}}{h_{g1} - h_{f2}} \tag{9.8}$$

It can be observed from Eq. (9.8) that the specific humidity ω_1 of the entering air–water vapor mixture can be measured by recording the inlet and outlet temperatures of the mixture.

EXAMPLE 9.4 A mixture of air and water vapor at 1 bar enters an adiabatic saturator at 40°C and leaves at 25°C. Calculate the specific humidity of the air–water vapor mixture.

SOLUTION At 25°C, P^s = 0.031 166 bar, h_f = 104.77 kJ/kg and h_g = 2547.3 kJ/kg

The air–water vapor mixture leaves the adiabatic saturator in the saturated state. Therefore, $P^s = p_w$ (or $\phi = 1$) at the exit of the device. Hence,

$$p_{w2} = 0.031\ 66 \text{ bar}$$

$$\omega_2 = \frac{0.622 p_{w2}}{P - p_{w2}} = \frac{0.622 \times 0.031\ 66}{1 - 0.031\ 66} = 0.0203 \frac{\text{kg water}}{\text{kg dry air}}$$

C_p of dry air = 1.005 kJ/kg K; At 40°C h_{g1} = 2574.4 kJ/kg

$$\omega_1 = \frac{C_{pa}(T_2 - T_1) + \omega_2 h_{fg2}}{h_{g1} - h_{f2}} = \frac{1.005(25 - 40) + 0.0203 \times (2547.3 - 104.77)}{2574.4 - 104.77}$$

or $\omega_1 = 0.013\ 97 \dfrac{\text{kg water}}{\text{kg dry air}}$

9.5 Psychrometer

A simple instrument called a psychrometer working on the principle of adiabatic saturation is used to measure the specific humidity of air–water vapor mixtures. A schematic of a sling psychrometer is shown in Fig. 9.2. It consists of a dry bulb thermometer and a wet bulb thermometer. The instrument can be whirled around in the air to provide sufficient flow of air–water vapor mixture over the two thermometers. The bulb of one of the thermometers, the dry bulb thermometer, is left uncovered and it reads the temperature of the air–water vapor mixture. This temperature is called the *dry bulb temperature*(T_{DB}). The bulb of the second thermometer, the wet bulb thermometer, is covered with a wet wick and the temperature read by this thermometer is called the *wet bulb temperature* (T_{WB}).

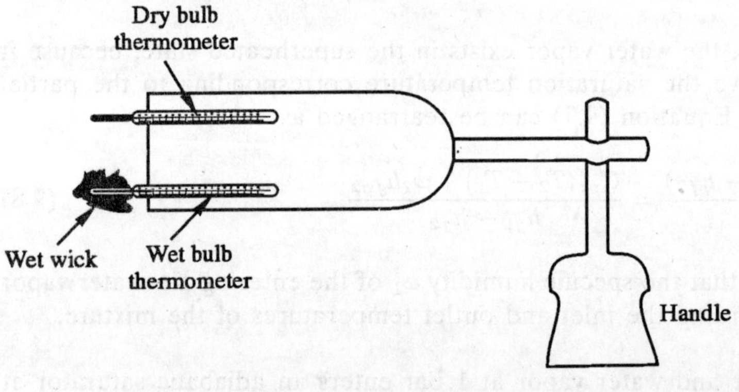

Fig. 9.2 *Schematic of a sling psychrometer.*

The dry bulb temperature is simply the temperature of the air–water vapor mixture. The wet bulb temperature depends on the rates of mass and energy transfer between the two phases, which in turn depend on the flow velocity of air. On the other hand, the adiabatic saturation temperature is a property of the air–water vapor mixture and is governed by the equilibrium between the two phases. It was experimentally observed that for air–water vapor mixtures, the adiabatic saturation temperature is approximately equal to the wet bulb temperature.

EXAMPLE 9.5 For a particular air–water vapor mixture at 101.325 kPa, a psychrometer reads 40°C as the dry bulb temperature and 25°C as the wet bulb temperature. Calculate the specific humidity and relative humidity of the air–water vapor mixture.

SOLUTION The wet bulb temperature T_{WB} is equal to the adiabatic saturation temperature T_2. Therefore at T_2 = 25°C, the air is saturated. That is,

$$p_{w2} = P^s(25°\text{C}) = 0.031\ 66 \text{ bar}$$

$$\omega_2 = \frac{0.622 p_{w2}}{P - p_{w2}} = \frac{0.622 \times 0.031\ 66}{1.013\ 25 - 0.031\ 66} = 0.02 \ \frac{\text{kg water}}{\text{kg dry air}}$$

$$h_{fg2} = 2442.53 \text{ kJ/kg}; \quad h_{g1} = 2574.4 \text{ kJ/kg};$$

$$h_{f2} = 104.77 \text{ kJ/kg}; \quad C_{pa} = 1.005 \text{ kJ/kg K}$$

$$\omega_1 = \frac{C_{pa}(T_2 - T_1) + \omega_2 h_{fg2}}{h_{g1} - h_{f2}} = \frac{1.005(25 - 40) + 0.02 \times 2442.53}{2574.4 - 104.77} = 0.0137 \ \frac{\text{kg water}}{\text{kg dry air}}$$

We know that

$$\omega_1 = \frac{0.662 p_{w1}}{P - p_{w1}} \quad \text{or} \quad 0.0137 = \frac{0.662 p_{w1}}{1.013\ 25 - p_{w1}}$$

$$p_{w1} = 0.0218 \text{ bar}$$

$$P^s(40°C) = 0.073\ 75 \text{ bar}$$

Therefore, $\phi_1 = \dfrac{p_{w1}}{P^s} = \dfrac{0.0218}{0.073\ 75} = 0.296$

That is, the relative humidity = 29.6%

9.6 Psychrometric Chart

The adiabatic saturation relation given by Eq. (9.8) can be used to calculate the specific humidity and relative humidity of an air–water vapor mixture. In the field of air-conditioning, an engineer extensively deals with air–water vapor mixtures and has to solve the adiabatic saturation relation quite often. To facilitate the use of the adiabatic saturation relation, its solution is presented in a convenient and ready-to-use graphical form. Such a graphical representation of the adiabatic saturation relation is called a psychrometric chart. Usually psychrometric charts are prepared for a given total pressure. The properties like dry bulb temperature, wet bulb temperature, specific humidity, relative humidity and the enthalpy of the mixture are shown on a single chart.

The psychrometric chart at a total pressure of 101.325 kPa, showing the important properties, is presented in Fig. 9.3. The enthalpy of the air–water vapor mixture is usually expressed on the basis of dry air. The enthalpy (h^*) of the air–water vapor mixture is given by

$$h^* = h_a + \omega h_w = C_{pa}T + \omega h_g \tag{9.9}$$

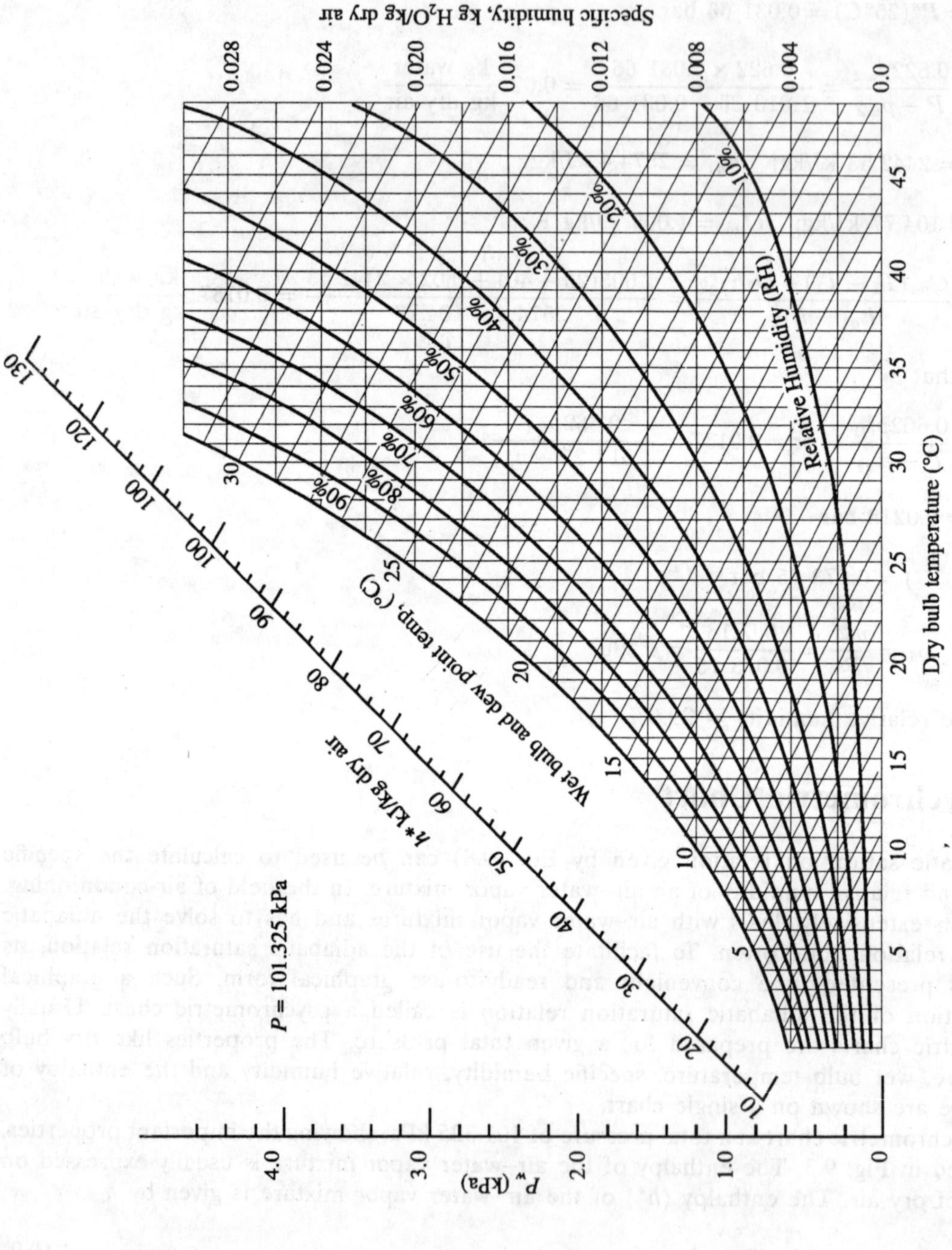

Fig. 9.3 *Psychrometric chart at a total pressure of 101.325 kPa.*

That is, (h^*) represents the enthalpy of 1 kg dry air and the enthalpy of the accompanying water vapor. The enthalpy of the air is measured relative to 0°C, while the enthalpy of water uses the reference state which is the same as that used in the steam tables. The adiabatic saturation Eq. (9.7) can be rearranged as

$$(C_{pa}T_1 + \omega_1 h_{g1}) + (\omega_2 - \omega_1)h_{f2} = (C_{pa}T_2 + \omega_2 h_{g2})$$

or $$h_1^* + (\omega_2 - \omega_1)h_{f2} = h_2^* \tag{9.10}$$

The value of the second term on the left-hand side of Eq. (9.10) is negligible compared to the other terms, and hence Eq. (9.10) can be written as

$$h_1^* = h_2^* \tag{9.11}$$

That is, during adiabatic saturation h^* remains constant. The value of h_2^* depends on the adiabatic saturation temperature only. Hence, the lines of constant h^* coincide with the lines of constant wet bulb temperature.

EXAMPLE 9.6 Calculate the specific humidity and relative humidity of an air–water vapor mixture if the dry bulb and wet bulb temperatures are 40°C and 25°C, respectively. Compare the results with the values read from the psychrometric chart.

SOLUTION The values of ω and ϕ have been calculated in Example 9.5.

$$\omega = 0.0137 \frac{\text{kg water}}{\text{kg dry air}} \quad \text{and} \quad \phi = 0.296$$

A schematic of the psychrometric chart is shown in Fig. 9.4. Locate the intersection of T_{DB} = 40°C and T_{WB} = 25°C as shown in Fig. 9.4, and read ω = 0.0137 kg water/kg dry air and ϕ = 30%.

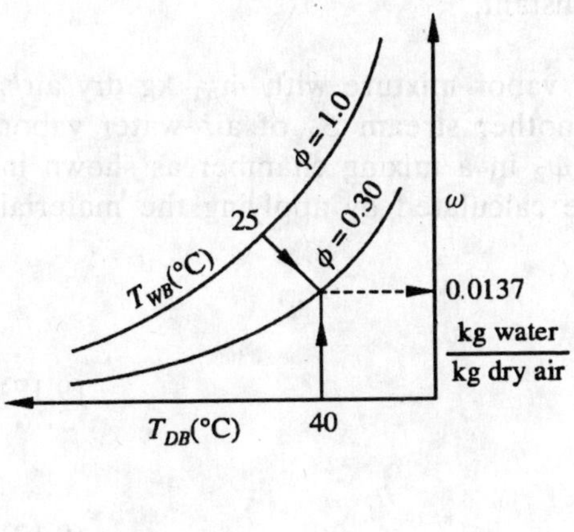

Fig. 9.4 Sketch for Example 9.6.

9.7 Air-Conditioning

The theory of air–water vapor mixtures is extensively used in the field of air-conditioning. Air-conditioning means conditioning the air to adjust the temperature and relative humidity of the air either for comfortable living or for the safe operation of electronic equipment and storage of food material. The common processes in the field of air-conditioning are heating, cooling, mixing of different streams, humidification and dehumidification.

Heating and Cooling Suppose an air–water vapor mixture which is initially at state 1 is cooled at constant pressure. It follows the path 1-2-3 as shown in Fig. 9.5.

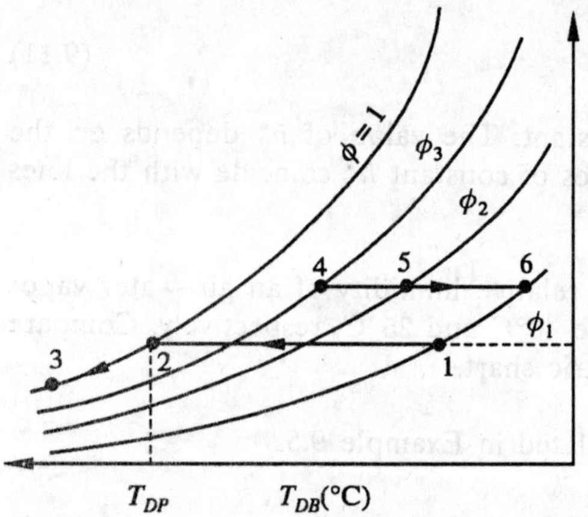

Fig. 9.5 *Psychrometric chart showing constant pressure cooling and heating processes.*

While cooling, the temperature of the air decreases till state 2 is reached and the specific humidity remains constant and the relative humidity keeps increasing. At state 2, the air gets saturated and reaches the dew point. Further cooling reduces the specific humidity, as water condenses, and the air follows the path 2-3 along the 100% relative humidity curve. If air at state 4 is heated at constant pressure, it follows the path 4-5-6 and the relative humidity decreases even though the specific humidity remains constant.

Adiabatic Mixing Suppose one stream S_1 of air–water vapor mixture with $\dot{m}_{a1}$ kg dry air/s with specific humidity ω_1 is adiabatically mixed with another stream S_2 of air–water vapor mixture with $\dot{m}_{a2}$ kg dry air/s having specific humidity ω_2 in a mixing chamber as shown in Fig. 9.6, the conditions of the resulting mixture can be calculated by applying the material and energy balances.

Mass balance for air,

$$\dot{m}_{a1} + \dot{m}_{a2} = \dot{m}_{a3} \tag{9.12}$$

Mass balance for water,

$$\dot{m}_{a1}\omega_1 + \dot{m}_{a2}\omega_2 = \dot{m}_{a3}\omega_3 \tag{9.13}$$

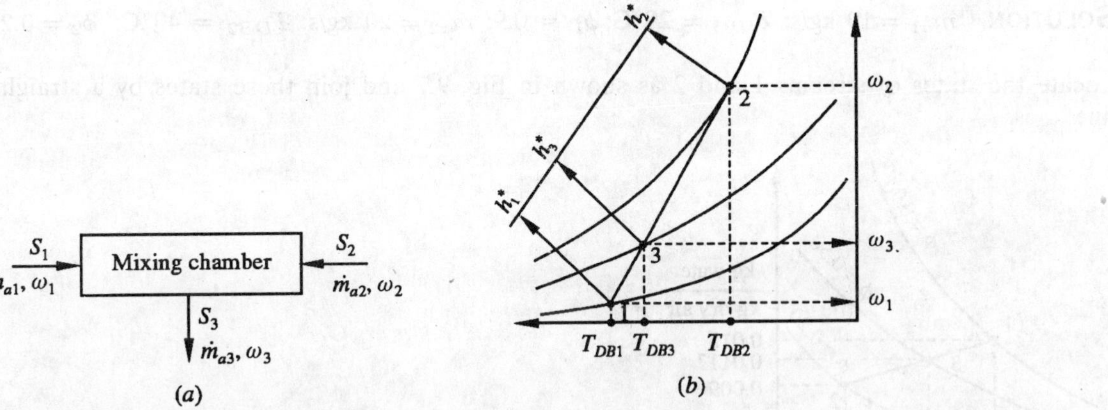

Fig. 9.6 (a) *Adiabatic mixing chamber.* (b) *Psychrometric chart showing adiabatic mixing processes.*

Energy balance,

$$\dot{m}_{a1}h_1^* + \dot{m}_{a2}h_2^* = \dot{m}_{a3}h_3^* \tag{9.14}$$

From Eqs. (9.12) and (9.13), we can write

$$\dot{m}_{a1}\omega_1 + \dot{m}_{a2}\omega_2 = (\dot{m}_{a1} + \dot{m}_{a2})\omega_3$$

or
$$\omega_3 = \frac{\dot{m}_{a1}\omega_1 + \dot{m}_{a2}\omega_2}{\dot{m}_{a1} + \dot{m}_{a2}} \tag{9.15}$$

or
$$\dot{m}_{a2}(\omega_2 - \omega_3) = \dot{m}_{a1}(\omega_3 - \omega_1)$$

$$\frac{\dot{m}_{a1}}{\dot{m}_{a2}} = \frac{\omega_2 - \omega_3}{\omega_3 - \omega_1} = \frac{\text{Distance } 2-3}{\text{Distance } 3-1} \tag{9.16}$$

From Eqs. (9.12) and (9.14), we obtain

$$h_3^* = \frac{\dot{m}_{a1}h_1^* + \dot{m}_{a2}h_2^*}{\dot{m}_{a1} + \dot{m}_{a2}} \tag{9.17}$$

or
$$\frac{\dot{m}_{a1}}{\dot{m}_{a2}} = \frac{h_2^* - h_3^*}{h_3^* - h_1^*} \tag{9.18}$$

That is, the resulting stream S_3 lies on the line joining the states of S_1 and S_2 as shown in Fig. 9.6(b).

EXAMPLE 9.7 If a stream of air–water vapor mixture at 20°C and $\phi = 0.9$ with $\dot{m}_a = 10$ kg/s is adiabatically mixed with another stream ($\dot{m}_a = 20$ kg/s) at 40°C and $\phi = 0.2$, calculate the temperature and relative humidity of the resulting mixture.

SOLUTION $\dot{m}_{a1} = 10$ kg/s; $T_{DB1} = 20°C$; $\phi_1 = 0.9$; $\dot{m}_{a2} = 20$ kg/s; $T_{DB2} = 40°C$; $\phi_2 = 0.2$

Locate the states of streams 1 and 2 as shown in Fig. 9.7 and join these states by a straight line.

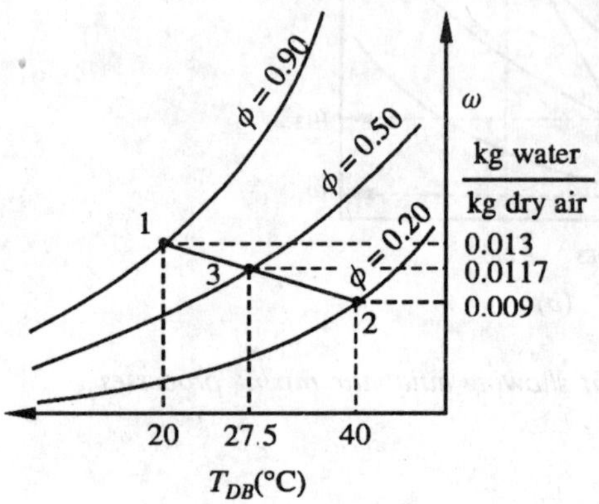

Fig. 9.7 Sketch for Example 9.7.

Then from the psychrometric chart read $\omega_2 = 0.009$ kg water/kg dry air and $\omega_1 = 0.013$ kg water/kg dry air.

$$\omega_3 = \frac{\dot{m}_{a1}\omega_1 + \dot{m}_{a2}\omega_2}{\dot{m}_{a1} + \dot{m}_{a2}} = \frac{10 \times 0.009 + 20 \times 0.013}{10 + 20} = 0.0117 \ \frac{\text{kg water}}{\text{kg dry air}}$$

Then on the straight line locate state 3 at $\omega_3 = 0.0117$ kg water/kg dry air, and read $\phi = 50\%$ and $T = 27.5°C$ at state 3.

Dehumidification In hot and humid climates, we feel the necessity of air-conditioning. That is, we need to control the humidity and temperature for comfortable living. This can be achieved by dehumidifying and cooling the air to the required conditions. Suppose the air–water vapor mixture is initially at state 1, having high relative humidity and dry bulb temperature, then following the path 1-A-B-2, shown in Fig. 9.8, the air can be conditioned.

The air is cooled at constant pressure following the path 1-A-B-2. During the process 1-A, the specific humidity of the air remains constant. At state A, the air gets saturated and on further cooling, the water vapor condenses and the air follows the path A-B. The condensed water at state B is removed and the saturated air at state B is heated at constant pressure along the path B-2 till the required dry bulb temperature is reached. In some commercial air-conditioning units, chilled water below the dew point temperature of the mixture is sprayed into the air to dehumidify the air as shown in Fig. 9.9.

The energy requirement for the refrigerating and heating units can be determined by applying material and energy balances to the individual units.

For the refrigerating unit Mass balance for water, $\dot{m}_a\omega_1 = \dot{m}_a\omega_2 + \dot{m}_c$

or

$$\dot{m}_c = \dot{m}_a(\omega_1 - \omega_2) \tag{9.19}$$

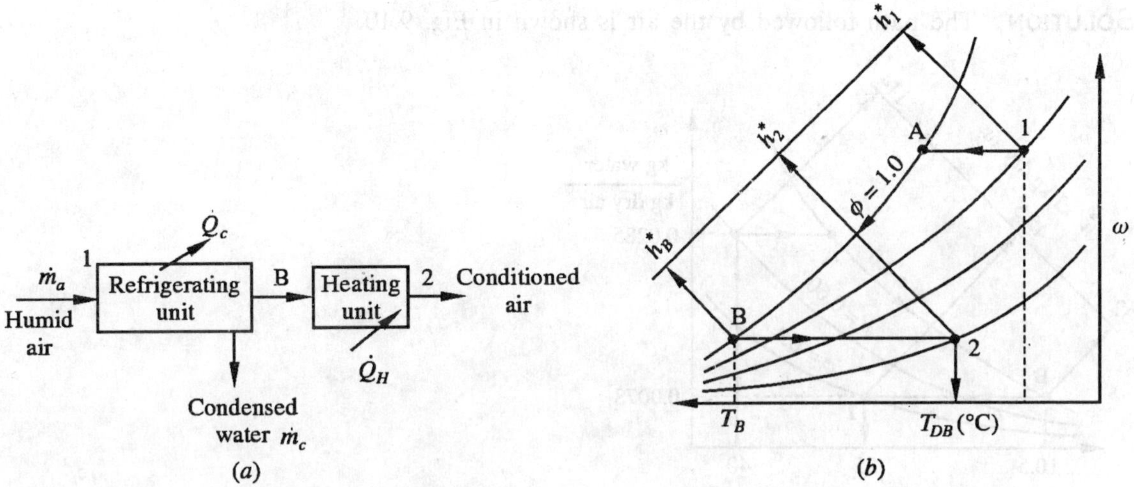

Fig. 9.8 *Psychrometric chart showing the path for air-conditioning.*

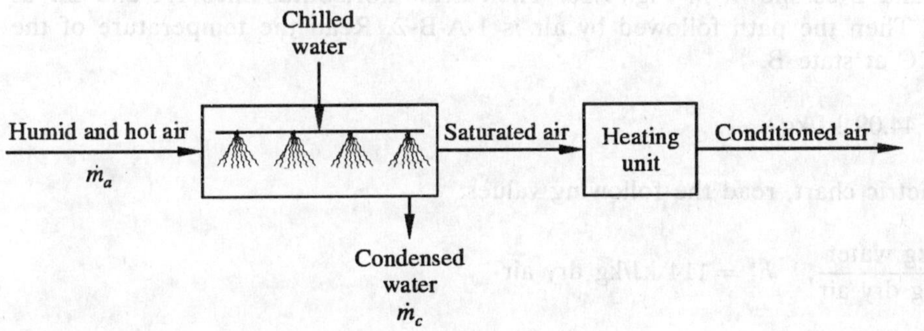

Fig. 9.9 *Schematic of a spray dehumidifier.*

Energy balance, $\dot{m}_a h_1^* = \dot{Q}_c + \dot{m}_c h_w + \dot{m}_a h_B^*$

or

$$Q_c = \dot{m}_a(h_1^* - h_B^*) - \dot{m}_c h_w \tag{9.20}$$

where h_w is the enthalpy of water vapor at T_B.

For the heating unit Energy balance, $\dot{m}_a h_B^* + \dot{Q}_H = \dot{m}_a h_2^*$

or

$$\dot{Q}_H = \dot{m}_a(h_2^* - h_B^*) \tag{9.21}$$

EXAMPLE 9.8 Humid air at 101.325 kPa and 40°C with a relative humidity of 60% is to be conditioned to 25°C and 40% relative humidity. If the required flow rate of air is 1 kg/s, estimate the capacities of the cooling and heating devices in the dehumidifier.

SOLUTION The path followed by the air is shown in Fig. 9.10.

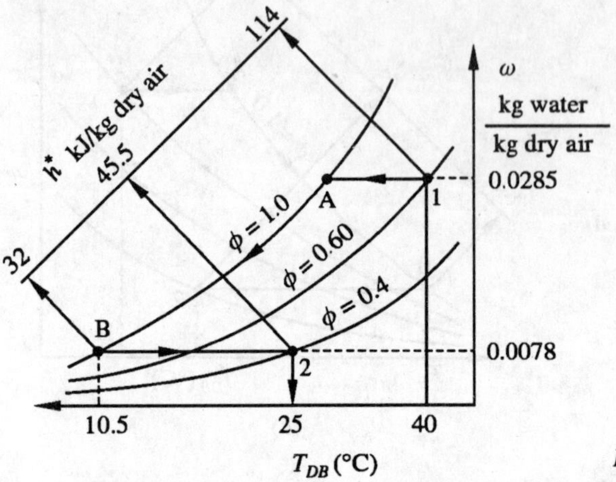

Fig. 9.10 Sketch for Example 9.8.

Locate states 1 and 2 as shown in Fig. 9.10. Then draw horizontal lines 1A and 2B as shown in Fig. 9.10. Then the path followed by air is 1-A-B-2. Read the temperature of the condensate as 10.5°C at state B.

$$h_f(10.5°\text{C}) = 44.09 \text{ kJ/kg}$$

From the psychrometric chart, read the following values:

$$\omega_1 = 0.0285 \frac{\text{kg water}}{\text{kg dry air}}; \quad h_1^* = 114 \text{ kJ/kg dry air}$$

$$\omega_2 = 0.0078 \frac{\text{kg water}}{\text{kg dry air}}; \quad h_2^* = 45.5 \text{ kJ/kg dry air}$$

$$h_B^* = 32 \text{ kJ/kg dry air}$$

Then the rate of cooling is given by

$$\dot{Q}_c = \dot{m}_a(h_1^* - h_B^*) - \dot{m}_c h_w = \dot{m}_a(h_1^* - h_B^*) - \dot{m}_a(\omega_1 - \omega_2)h_w$$

$$= (114 - 32) - (0.0285 - 0.0078) \times 44.09 = 81.09 \text{ kW}$$

The rate of heating is given by

$$\dot{Q}_H = \dot{m}_a(h_2^* - h_B^*) = 45.5 - 32 = 13.5 \text{ kW}$$

Cooling by Humidification If an unsaturated air–water vapor mixture is passed through pads which are kept wet, the mixture leaves in a saturated or near saturated state. The energy required for the evaporation of water to increase the humidity of air comes from the air and hence air gets cooled. The process that occurs is similar to adiabatic saturation, except that the air may not leave the unit in the saturated state. This process is used in desert coolers which are in wide use in hot and dry climates for cooling houses. The device is also known as an evaporative cooler. A schematic of an evaporative cooler and the process that occurs during the passage of air through the wet porous pads is shown in Fig. 9.11.

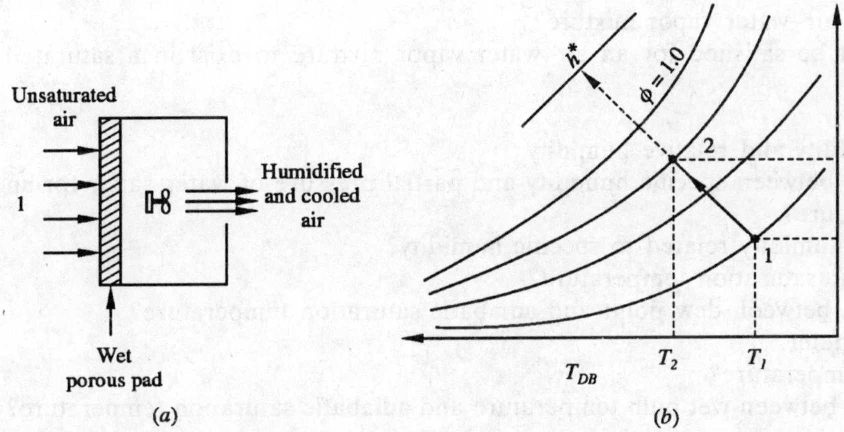

Fig. 9.11 (a) Schematic of a desert cooler or evaporative cooler. (b) Psychrometric chart showing the process of cooling by humidification.

EXAMPLE 9.9 On a hot summer day, the atmospheric air is at 40°C with 20% relative humidity. If the exhaust fan in a desert cooler can handle 1 kg of dry air/s, determine the temperature to which the air can be cooled and the amount of water needed to operate the cooler for one hour.

SOLUTION Locate the initial state 1 of the entering unsaturated air as shown in Fig. 9.12 and then follow the constant wet bulb temperature line till it intersects the $\phi = 1$ curve. Then read the temperature as 21.8°C.

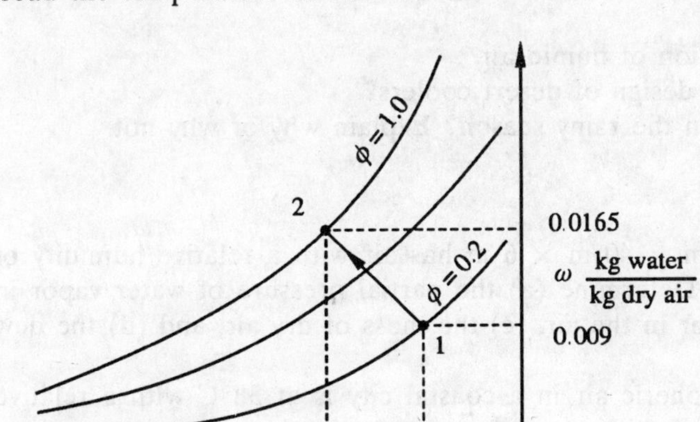

Fig. 9.12 Sketch for Example 9.9.

From the psychrometric chart, read $\omega_1 = 0.009$ kg water/kg dry air and $\omega_2 = 0.0165$ kg water/kg dry air. Mass of water required $= \dot{m}_a \times (\omega_2 - \omega_1) = 1 \times (0.0165 - 0.009) = 0.0075$ kg/s = 27 kg/hour

Review Questions

1. What is a saturated air–water vapor mixture?
2. What criterion must be satisfied for an air–water vapor mixture to exist in a saturated state?
3. What is dew point?
4. Define specific humidity and relative humidity.
5. What is the relation between specific humidity and partial pressure of water vapor for an air–water vapor mixture?
6. How is the relative humidity related to specific humidity?
7. What is the adiabatic saturation temperature?
8. What is the relation between dew point and adiabatic saturation temperature?
9. Describe a psychrometer.
10. What is wet bulb temperature?
11. What is the relation between wet bulb temperature and adiabatic saturation temperature?
12. Sketch a psychrometric chart and explain the salient features.
13. What is the relation between dry bulb temperature, wet bulb temperature, dew point and adiabatic saturation temperature for a saturated air–water vapor mixture?
14. How do you find the state of the resulting mixture if two streams of air–water vapor mixture with different temperatures and specific humidities are adiabatically mixed?
15. Do the specific humidity and relative humidity of an unsaturated air–water vapor mixture change if it is cooled at constant pressure to a final temperature which is above the dew point?
16. Does the specific humidity of an unsaturated air–water vapor mixture change if it is cooled at constant pressure to a final temperature which is below the dew point?
17. Define the enthalpy of an air–water vapor mixture.
18. What are the reference states used while calculating the enthalpy of air–water vapor mixtures?
19. Suggest a method for dehumidification of humid air.
20. Which principle is employed in the design of desert coolers?
21. Will the desert cooler be effective in the rainy season? Explain why or why not.

Problems

9.1 An auditorium of dimensions 40 m × 40 m × 6 m has air with a relative humidity of 60% at 1 atm pressure and 30°C. Determine (a) the partial pressure of water vapor in the air, (b) the mass of water vapor in the air, (c) the mass of dry air, and (d) the dew point.

9.2 On a hot summer day, the atmospheric air in a coastal city is at 38°C with a relative humidity of 80%. Calculate the dew point.

9.3 The partial pressure of water vapor in an air–water vapor mixture at 1 atm pressure and 40°C is 0.032 bar. Calculate the specific humidity and relative humidity of the air.

9.4 On a particular day, the partial pressure of water vapor in the atmospheric air is 0.042 bar at 36°C. Calculate the dew point of the air.

9.5 A mixture of air and water vapor enters an adiabatic saturator at 35°C and 1 atm pressure and leaves at 25°C. Calculate the specific humidity and relative humidity of the air.

9.6 On a particular day, a psychrometer reads 35°C and 26°C as dry bulb and wet bulb temperatures, respectively. If the pressure is 101.325 kPa, determine the specific humidity and relative humidity using the adiabatic saturation relation.

9.7 Rework Problem 9.6 using the psychrometric chart.

9.8 A psychrometer is used to measure the dry bulb and wet bulb temperatures of atmospheric air on a particular day and it is found that T_{DB} = 40°C and T_{WB} = 30°C. Using the psychrometric chart, determine the specific humidity, relative humidity and enthalpy of air.

9.9 An air-conditioner delivers 1 m^3/s of air at 5°C which is saturated. This air is mixed with 5 m^3/s of air at 40°C with 50% relative humidity. Determine the temperature and specific humidity of the resultant mixture.

9.10 Two streams S_1 and S_2 of air are mixed as shown in Fig. 9.13 to obtain a mixture at 25°C with a relative humidity of 40%. Determine the relative humidity and the flow rate of stream S_1.

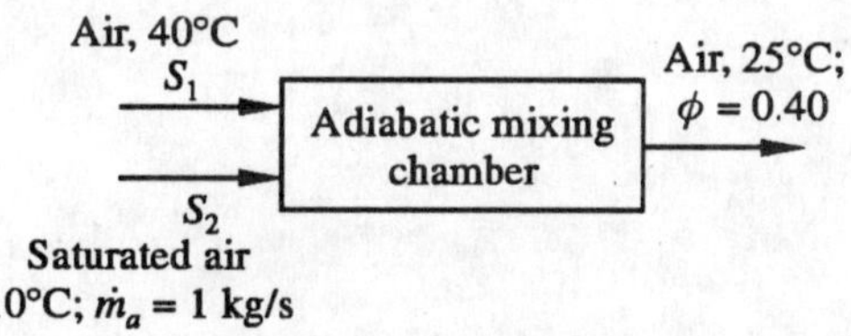

Fig. 9.13 Sketch for Problem 9.10.

9.11 Two air-conditioners blow air into a common duct. The first air-conditioner delivers air at 10°C and ϕ = 0.90 at a rate of 1 kg dry air/s, while the second air-conditioner blows air at 20°C and ϕ = 0.70 at a rate of 2 kg dry air/s. Determine the temperature and relative humidity of the mixture.

9.12 If air at 40°C with ϕ = 0.40 is cooled at a constant pressure of 1 atm to 32°C, calculate the specific humidity and relative humidity of air in the final state.

9.13 An air–water vapor mixture at 40°C with ϕ = 0.5 is cooled to 10°C at a constant pressure of 1 atm. Calculate the specific humidity of the air in the final state. Also calculate the mass of water condensed if the original air–water vapor mixture contains 10 kg dry air.

9.14 Moist air enters a cooling device at 40°C with 60% relative humidity. The air leaves the device as saturated air at 20°C. The cooling device is operating at a constant pressure of 1 atm. If the flow rate of air through the device is 10 kg dry air/s, calculate the cooling capacity of the device and the rate at which water is condensing in the device.

9.15 It is required to dry a material by passing air over it. In a particular drier, air at 30°C with 20% relative humidity enters, and leaves at 40°C with 60% relative humidity. Determine the air flow rate if 10 kg of water is to be evaporated from the material in one hour.

9.16 Atmospheric air at 40°C with 50% relative humidity is to be conditioned to 25°C with 40% relative humidity. Suggest a method and estimate the cooling capacity and heating capacity of the unit if the flow rate of air through the device is 10 kg dry air/s.

9.17 On a hot summer day, the atmospheric air is at 45°C with 20% relative humidity. It is proposed to use a desert cooler to cool a house. Estimate the minimum temperature to which the air can be cooled in a perfectly designed desert cooler. Suppose the fan of the desert cooler can draw 2 kg dry air/s, estimate the mass of water required to operate the desert cooler for 3 hours.

9.18 Cold air is to be blown into an auditorium at a rate of 20 kg dry air/s. For this purpose a desert cooler which takes in atmospheric air at 40°C with 20% relative humidity is used. The air leaves the cooler with 80% relative humidity. Estimate the temperature of the air leaving the cooler and the quantity of water needed to operate the cooler for 3 hours.

9.19 A desert cooler is used to blow cold air into a room. If the atmospheric air is at 37°C with 60% relative humidity and the air leaves the cooler with 90% relative humidity, estimate the temperature of the air leaving the cooler.

9.20 The partial pressure of water vapor in a gas mixture is 20 kPa while the mixture is at 1 bar. If the mixture is cooled at constant pressure, estimate the temperature at which water starts condensing. If the temperature of the mixture is reduced by 10°C below the temperature at which condensation begins, estimate the partial pressure of water vapor in the final mixture.

APPENDICES

Table A1 Critical constants of some selected substances

Compound		Critical temperature T_c (K)	Critical pressure P_c (bar)	Critical volume v_c (m^3/kmol)
Inorganic compounds				
Ammonia	NH_3	405.50	112.77	0.0725
Argon	A	151.20	48.64	0.075
Bromine	Br_2	584.00	103.35	0.114
Carbon dioxide	CO_2	304.20	73.87	0.094
Carbon disulphide	CS_2	552.00	79.03	0.170
Carbon monoxide	CO	133.00	34.96	0.093
Chlorine	Cl_2	417.00	77.11	0.124
Helium	He	5.26	22.90	0.058
Hydrogen	H_2	33.30	129.70	0.065
Hydrogen chloride	HCl	324.60	82.58	0.087
Hydrogen sulphide	H_2S	373.60	90.08	0.098
Neon	Ne	44.50	27.26	0.0417
Nitrogen	N_2	126.20	33.94	0.090
Nitric oxide	NO	179.20	65.86	0.058
Nitrous oxide	N_2O	309.70	72.65	0.0963
Oxygen	O_2	154.40	50.36	0.074
Sulphur dioxide	SO_2	430.70	78.83	0.122
Sulphur trioxide	SO_3	491.40	84.91	0.126
Water	H_2O	647.30	221.20	0.056
Organic compounds				
Acetic acid	CH_3COOH	594.80	57.86	0.171
Acetone	CH_3COCH_3	508.70	47.22	0.213
Acetylene	C_2H_2	309.50	62.42	0.113
Aniline	C_6H_7N	698.80	52.99	0.274
Benzene	C_6H_6	562.10	49.24	0.260
Butane (iso)	C_4H_{10}	408.00	36.48	0.263
Butane (n)	C_4H_{10}	425.20	37.97	0.255
Butene (1)	C_4H_8	419.60	40.23	0.240
Carbon tetrachloride	CCl_4	556.40	45.60	0.276
Chlorobenzene	C_6H_5Cl	632.40	45.19	0.308
Chlorodifluoromethane	$CHClF_2$	369.60	49.14	0.165

Continues

Table A1 Continued

Compound		Critical temperature T_c (K)	Critical pressure P_c (bar)	Critical volume v_c (m^3/kmol)
Chloroform	$CHCl_3$	536.60	54.72	0.240
Cyclopentane	C_5H_{10}	511.76	21.08	0.260
Diethyl ether	$C_2H_5OC_2H_5$	467.00	39.62	0.281
Ethane	C_2H_6	305.43	48.84	0.148
Ethylamine	$C_2H_5NH_2$	456.40	56.24	0.181
Ethyl acetate	$C_2H_5COOCH_3$	523.30	38.30	0.286
Ethyl alcohol	C_2H_5OH	516.30	68.83	0.167
Ethylene	C_2H_4	283.10	51.17	0.124
Ethylene oxide	C_2H_4O	468.00	71.94	0.138
Freon-11	CCl_3F	471.20	43.77	0.248
Freon-12	CCl_2F_2	384.70	40.12	0.218
Freon-13	$CClF_3$	302.00	38.71	0.180
Freon-21	$CHCl_2F$	451.70	51.68	0.197
Freon-113	$C_2Cl_3F_3$	487.30	34.15	0.325
Heptane (n)	C_7H_{16}	540.16	27.37	0.426
Hexane (n)	C_6H_{14}	507.90	30.32	0.368
Methane	CH_4	190.70	46.41	0.099
Methyl acetate	CH_3COOCH_3	506.90	46.91	0.228
Methyl chloride	CH_3Cl	416.30	66.77	0.143
Methyl alcohol	CH_3OH	513.20	75.54	0.118
Octane (n)	C_8H_{18}	569.40	24.97	0.486
Pentane (n)	C_5H_{12}	469.80	33.75	0.311
Propane	C_3H_8	369.90	42.57	0.200
Propyl alcohol (n)	C_3H_7OH	537.30	50.87	0.220
Propyl alcohol (iso)	C_3H_7OH	508.80	53.70	0.219
Propylene	C_3H_6	365.10	46.00	0.181
Toluene	$C_6H_5CH_3$	594.00	42.15	0.320

Table A2 Saturated steam : Temperature table

Temperature t (°C)	Pressure P (bar)	Specific volume v_f (m^3/kg)	Specific volume v_g (m^3/kg)	Specific enthalpy h_f (kJ/kg)	Specific enthalpy h_g (kJ/kg)	Specific entropy s_f (kJ/kg K)	Specific entropy s_g (kJ/kg K)
0	0.006 108	0.001 000 2	206.3000	– 0.04	2501.6	– 0.0002	9.1577
2	0.007 055	0.001 000 1	179.9000	8.39	2505.2	0.0306	9.1047
4	0.008 129	0.001 000 0	157.3000	16.80	2508.9	0.0611	9.0526
6	0.009 345	0.001 000 0	137.8000	25.21	2512.6	0.0913	9.0015
8	0.010 720	0.001 000 1	121.0000	33.60	2516.2	0.1213	8.9513
10	0.012 270	0.001 000 3	106.4000	41.99	2519.9	0.1510	8.9020
15	0.017 039	0.001 000 8	77.9800	62.94	2529.1	0.2243	8.7826
20	0.023 37	0.001 001 7	57.8400	83.86	2538.2	0.2963	8.6684
25	0.031 66	0.001 002 9	43.4000	104.77	2547.3	0.3670	8.5592
30	0.042 41	0.001 004 3	32.9300	125.66	2556.4	0.4365	8.4546
35	0.056 22	0.001 006 0	25.2400	146.56	2565.4	0.5049	8.3543
40	0.073 75	0.001 007 8	19.5500	167.45	2574.4	0.5721	8.2583
45	0.095 82	0.001 009 9	15.2800	188.35	2583.3	0.6383	8.1661
50	0.123 35	0.001 012 1	12.0500	209.26	2592.2	0.7035	8.0776
55	0.157 41	0.001 014 5	9.5790	230.17	2601.0	0.7677	7.9926
60	0.199 20	0.001 017 1	7.6790	251.09	2609.7	0.8310	7.9108
65	0.2501	0.001 019 9	6.2020	272.02	2618.4	0.8933	7.8322
70	0.3116	0.001 022 8	5.0460	292.97	2626.9	0.9548	7.7565
75	0.3855	0.001 025 9	4.1340	313.94	2635.4	1.0154	7.6835
80	0.4736	0.001 029 2	3.4090	334.92	2643.8	1.0753	7.6132
85	0.5780	0.001 032 6	2.8290	355.92	2652.0	1.1343	7.5454
90	0.7011	0.001 036 1	2.3610	376.94	2660.1	1.1925	7.4799
95	0.8453	0.001 039 9	1.9820	397.99	2668.1	1.2501	7.4166
100	1.0133	0.001 043 7	1.6730	419.06	2676.0	1.3069	7.3554
105	1.2080	0.001 047 7	1.4190	440.17	2683.7	1.3630	7.2962
110	1.4327	0.001 051 9	1.2100	461.32	2691.3	1.4185	7.2388
115	1.6906	0.001 056 2	1.0360	482.50	2698.7	1.4733	7.1832
120	1.9854	0.001 060 6	0.8915	503.72	2706.0	1.5276	7.1293
125	2.3210	0.001 065 2	0.7702	524.99	2713.0	1.5813	7.0769
130	2.7013	0.001 070 0	0.6681	546.31	2719.9	1.6344	7.0261
135	3.131	0.001 075 0	0.5818	567.68	2726.6	1.6869	6.9766
140	3.614	0.001 080 1	0.5085	589.10	2733.1	1.7390	6.9284
145	4.155	0.001 085 3	0.4460	610.60	2739.3	1.7906	6.8815

Continues

Table A2 Continued

Temperature t (°C)	Pressure P (bar)	Specific volume v_f (m^3/kg)	v_g (m^3/kg)	Specific enthalpy h_f (kJ/kg)	h_g (kJ/kg)	Specific entropy s_f (kJ/kg K)	s_g (kJ/kg K)
150	4.760	0.001 090 8	0.3924	632.15	2745.4	1.8416	6.8358
155	5.433	0.001 096 4	0.3464	653.78	2751.2	1.8923	6.7911
160	6.181	0.001 102 2	0.3068	675.47	2756.7	1.9425	6.7475
165	7.008	0.001 108 2	0.2724	697.25	2762.0	1.9923	6.7048
170	7.920	0.001 114 5	0.2426	719.12	2667.1	2.0416	6.6630
175	8.924	0.001 120 9	0.2165	741.07	2771.8	2.0906	6.6221
180	10.027	0.001 127 5	0.1938	763.12	2776.3	2.1393	6.5819
185	11.233	0.001 134 4	0.1739	785.26	2780.4	2.1876	6.5424
190	12.551	0.001 141 5	0.1563	807.52	2784.3	2.2356	6.5036
195	13.987	0.001 148 9	0.1408	829.88	2787.8	2.2830	6.4654
200	15.549	0.001 156 5	0.1272	852.37	2790.9	2.3307	6.4278
205	17.243	0.001 164 4	0.1150	874.99	2793.8	2.3778	6.3906
210	19.077	0.001 172 6	0.1042	897.74	2796.2	2.4247	6.3539
215	21.060	0.001 181 1	0.094 63	920.63	2798.3	2.4713	6.3176
220	23.198	0.001 190 0	0.086 04	943.67	2799.9	2.5178	6.2817
225	25.501	0.001 199 2	0.078 35	966.89	2801.2	2.5641	6.2461
230	27.976	0.001·208 7	0.071 45	990.26	2802.0	2.6102	6.2107
235	30.632	0.001 218 7	0.065 25	1013.80	2802.3	2.6562	6.1756
240	33.478	0.001 229 1	0.059 65	1037.60	2802.2	2.7020	6.1406
245	36.523	0.001 239 9	0.054 61	1061.60	2801.6	2.7478	6.1057
250	39.776	0.001 251 3	0.050 04	1085.80	2800.4	2.7935	6.0708
255	43.246	0.001 263 2	0.045 90	1110.20	2798.7	2.8392	6.0359
260	46.943	0.001 275 6	0.042 13	1134.90	2796.4	2.8848	6.0010
270	55.058	0.001 302 5	0.035 59	1185.20	2789.9	2.9763	5.9304
280	64.202	0.001 332 4	0.030 13	1236.80	2780.4	3.0683	5.8586
290	74.461	0.001 365 9	0.025 54	1290.00	2767.6	3.1611	5.7848
300	85.927	0.001 404 1	0.021 65	1345.00	2751.0	3.2552	5.7081
310	98.700	0.001 448 0	0.018 33	1402.40	2730.0	3.3512	5.6278
320	112.890	0.001 499 5	0.015 48	1462.60	2703.7	3.4500	5.5423
330	128.630	0.001 561 5	0.012 99	1526.50	2670.2	3.5528	5.4490
340	146.050	0.001 638 7	0.010 78	1595.50	2626.2	3.6616	5.3427
350	165.350	0.001 741 1	0.008 799	1671.90	2567.7	3.7800	5.2177
360	186.750	0.001 895 9	0.006 940	1764.20	2485.4	3.9210	5.0600
370	210.540	0.002 213 6	0.004 973	1890.20	2342.8	4.1108	4.8114
374.15	221.200	0.003 17	0.003 17	2107.40	2107.4	4.4429	4.4429

Table A3 Saturated steam : Pressure table

Pressure P (bar)	Temperature t (°C)	Specific volume v_f (m^3/kg)	Specific volume v_g (m^3/kg)	Specific enthalpy h_f (kJ/kg)	Specific enthalpy h_g (kJ/kg)	Specific entropy s_f (kJ/kg K)	Specific entropy s_g (kJ/kg K)
0.01	6.98	0.001 000 1	129.20	29.34	2514.4	0.1060	8.9767
0.05	32.90	0.001 005 2	28.19	137.77	2561.6	0.4763	8.3960
0.10	45.83	0.001 010 2	14.67	191.83	2584.8	0.6493	8.1511
0.15	54.00	0.001 014 0	10.02	225.97	2599.2	0.7549	8.0093
0.20	60.09	0.001 017 2	7.650	251.45	2609.9	0.8321	7.9094
0.3	69.12	0.001 022 3	5.229	289.30	2625.4	0.9441	7.7695
0.4	75.89	0.001 026 5	3.993	317.65	2636.9	1.0261	7.6709
0.5	81.35	0.001 030 1	3.240	340.56	2646.0	1.0912	7.5947
0.6	85.95	0.001 033 3	2.732	359.93	2653.6	1.1454	7.5327
0.7	89.96	0.001 036 1	2.365	376.77	2660.1	1.1921	7.4804
0.8	93.51	0.001 038 7	2.087	391.72	2665.8	1.2330	7.4352
0.9	96.71	0.001 041 2	1.869	405.21	2670.9	1.2696	7.3954
1.0	99.63	0.001 043 4	1.694	417.54	2675.4	1.3027	7.3598
1.5	111.37	0.001 053 0	1.159	467.13	2693.4	1.4336	7.2234
2.0	120.23	0.001 060 8	0.8854	504.70	2706.3	1.5301	7.1268
2.5	127.43	0.001 067 5	0.7184	535.34	2716.4	1.6071	7.0520
3.0	133.54	0.001 073 5	0.6056	561.43	2724.7	1.6716	6.9909
3.5	138.87	0.001 078 9	0.5240	584.27	2731.6	1.7273	6.9392
4.0	143.62	0.001 083 9	0.4622	604.67	2737.6	1.7764	6.8943
5.0	151.84	0.001 092 8	0.3747	640.12	2747.5	1.8604	6.8192
6.0	158.84	0.001 100 9	0.3155	670.42	2755.5	1.9308	6.7575
7.0	164.96	0.001 108 9	0.2727	697.06	2762.0	1.9918	6.7052
8.0	170.41	0.001 115 0	0.2403	720.94	2767.5	2.0457	6.6192
9.0	175.36	0.001 121 3	0.2148	742.64	2772.1	2.0941	6.6192
10.0	179.88	0.001 127.4	0.1943	762.61	2776.2	2.1382	6.5862
11.0	184.07	0.001 133 1	0.1774	781.13	2779.7	2.1786	6.5497
12.0	187.96	0.001 138 6	0.1632	798.43	2782.7	2.2161	6.5194
13.0	191.61	0.001 143 8	0.1511	814.70	2785.4	2.2510	6.4913
14.0	195.04	0.001 148 9	0.1407	830.08	2787.8	2.2837	6.4651
15.0	198.29	0.001 153 9	0.1317	844.67	2789.9	2.3145	6.4406
16.0	201.37	0.001 158 6	0.1237	858.56	2791.7	2.3436	6.4175
17.0	204.31	0.001 163 3	0.1166	871.84	2793.4	2.3713	6.3957
18.0	207.11	0.001 167 8	0.1103	884.58	2794.8	2.3976	6.3751
19.0	209.80	0.001 172 3	0.1047	896.81	2796.1	2.4228	6.3554
20.0	212.37	0.001 176 6	0.099 54	908.59	2797.2	2.4469	6.3367

Continues

Table A3 Continued

Pressure P (bar)	Temperature t (°C)	Specific volume v_f (m^3/kg)	Specific volume v_g (m^3/kg)	Specific enthalpy h_f (kJ/kg)	Specific enthalpy h_g (kJ/kg)	Specific entropy s_f (kJ/kg K)	Specific entropy s_g (kJ/kg K)
25	223.94	0.001 197 2	0.079 91	961.96	2800.9	2.5543	6.2536
30	233.84	0.001 216 3	0.066 36	1008.40	2802.3	2.6455	6.1837
35	242.54	0.001 234 5	0.057 03	1049.80	2802.0	2.7253	6.1228
40	250.33	0.001 252 1	0.049 75	1087.40	2800.3	2.7965	6.0685
45	257.41	0.001 269 1	0.044 04	1122.10	2797.7	2.8612	6.0191
50	263.91	0.001 285 8	0.039 43	1154.50	2794.2	2.9206	5.9735
60	275.55	0.001 318 7	0.032 44	1213.70	2785.0	3.0273	5.8908
70	285.79	0.001 351 3	0.027 37	1267.40	2773.5	3.1219	5.8162
80	294.97	0.001 384 2	0.023 53	1317.10	2759.9	3.2076	5.7471
90	303.31	0.001 417 9	0.020 50	1363.70	2744.6	3.2867	5.6820
100	310.96	0.001 452 6	0.018 04	1408.00	2727.7	3.3605	5.6198
120	324.65	0.001 526 8	0.014 28	1491.80	2689.2	3.4972	5.5002
140	336.64	0.001 610 6	0.011 50	1571.60	2642.4	3.6242	5.3803
160	347.33	0.001 710 3	0.009 308	1650.50	2584.9	3.7471	5.2531
180	356.96	0.001 839 9	0.007 498	1734.80	2513.9	3.8765	5.1128
200	365.70	0.002 037 0	0.005 877	1826.50	2418.4	4.0149	4.9412
220	373.69	0.002 671 4	0.003 728	2011.10	2195.6	4.2947	4.5799
221.20	374.15	0.003 17	0.003 17	2107.4	2107.40	4.4429	4.4429

***Table A4** Superheated Steam*

v (m^3/kg): h (kJ/kg); s (kJ/kg K)

Pressure (bar) (Saturation temperature)		Temperature (°C)							
		100	200	300	400	500	600	700	800
0.2	v	8.585	10.507	13.219	15.529	17.838	20.146	22.455	24.762
(60.09°C)	h	2686.3	2879.2	3076.4	3279.4	3489.0	3705.4	3928.7	4158.7
	s	8.1261	8.5839	8.9618	9.2882	9.5784	9.8416	10.0836	10.3085
0.4	v	4.2792	5.4478	6.6065	7.7625	8.9176	10.072	11.227	12.381
(75.89°C)	h	2683.8	2878.2	3075.9	3279.1	3488.8	3705.3	3928.6	4158.6
	s	7.8009	8.2625	8.6413	8.9680	9.2583	9.5216	9.7636	9.9885
0.6	v	2.8440	3.6280	4.4020	5.1740	5.9440	6.7140	7.4840	8.2540
(85.95°C)	h	2681.3	2877.3	3075.4	3278.8	3488.6	3705.1	3928.5	4158.5
	s	7.6085	8.0738	8.4536	8.7806	9.0710	9.3343	9.5764	9.8013
0.8	v	2.1260	2.7180	3.3000	3.8790	4.4570	5.0350	5.6130	6.1900
(93.51°C)	h	2678.8	2876.3	3075.0	3278.5	3488.4	3705.0	3928.4	4158.4
	s	7.4703	7.9395	8.3202	8.6475	8.9380	9.2014	9.4436	9.6685
1.0	v	1.6960	2.1720	2.6390	3.1020	3.5650	4.0280	4.4900	4.9520
(99.63°C)	h	2676.2	2875.4	3074.5	3278.2	3488.1	3704.8	3928.2	4158.3
	s	7.3618	7.8349	8.2166	8.5442	8.8348	9.0982	9.3405	9.5654
2	v		1.0804	1.3162	1.5492	1.7812	2.0129	2.2420	2.4754
(120.23°C)	h		2870.5	3072.1	3276.7	3487.0	3704.0	3927.6	4157.8
	s		7.5072	7.8937	8.2226	8.5139	8.7776	9.0201	9.2452
3	v		0.7164	0.8753	1.0314	1.1865	1.3412	1.4957	1.6499
(133.54°C)	h		2865.5	3069.7	3275.2	3486.0	3703.2	3927.0	4157.3
	s		7.3119	7.7034	8.0338	8.3257	8.5868	8.8325	9.0577
4	v		0.5343	0.6549	0.7725	0.8892	1.0054	1.1214	1.2372
(143.62°C)	h		2860.4	3067.2	3273.6	3484.9	3702.3	3926.4	4156.9
	s		7.1708	7.5675	7.8994	8.1919	8.4563	8.6992	8.9246
5	v		0.4250	0.5226	0.6172	0.7108	0.8039	0.8968	0.9896
(151.84°C)	h		2855.1	3064.8	3272.1	3483.8	3701.5	3925.8	4156.4
	s		7.0592	7.4614	7.7948	8.0879	8.3526	8.5957	8.8213
6	v		0.3520	0.4344	0.5136	0.5918	0.6696	0.7471	0.8245
(158.84°C)	h		2849.7	3062.3	3270.6	3482.7	3700.7	3925.1	4155.9
	s		6.9662	7.3740	7.7090	8.0027	8.2678	8.5111	8.7368
7	v		0.2999	0.3714	0.4396	0.5069	0.5737	0.6402	0.7066
(164.96°C)	h		2844.2	3059.8	3269.8	3481.6	3699.9	3924.5	4155.5
	s		6.8859	7.2997	7.6362	7.9305	8.1959	8.4395	8.6653
8	v		0.2608	0.3241	0.3842	0.4432	0.5017	0.5600	0.6181
(170.41°C)	h		2838.6	3057.3	3267.5	3480.5	3699.1	3923.9	4115.0
	s		6.8148	7.2348	7.5729	7.8678	8.1336	8.3773	8.6033

Continues

Table A4 *Continued*

Pressure (bar)		Temperature (°C)						
(Saturation temperature)		200	300	400	500	600	700	800
9	*v*	0.2303	0.2874	0.3410	0.3936	0.4458	0.4976	0.5493
(175.36°C)	*h*	2832.7	3054.7	3266.0	3479.4	3698.2	3923.3	4154.8
	s	6.7508	7.1771	7.5169	7.8124	8.0785	8.3225	8.5486
10	*v*	0.2059	0.2580	0.3065	0.3540	0.4010	0.4477	0.4943
(179.88°C)	*h*	2826.8	3052.1	3264.4	3478.3	3697.4	3922.7	4154.1
	s	6.6922	7.1251	7.4665	7.7627	8.0292	8.2734	8.4997
11	*v*	0.1859	0.2339	0.2782	0.3215	0.3643	0.4069	0.4493
(184.07°C)	*h*	2820.7	3049.6	3262.9	3477.2	3696.6	3922.0	4153.6
	s	6.6379	7.0778	7.4209	7.7177	7.9845	8.2289	8.4553
12	*v*	0.1692	0.2139	0.2547	0.2945	0.3338	0.3729	0.4118
(187.96°C)	*h*	2814.4	3046.9	3261.3	3476.1	3695.8	3921.4	4153.1
	s	6.5872	7.0342	7.3790	7.6765	7.9436	8.1882	8.4148
13	*v*	0.1551	0.1969	0.2348	0.2716	0.3080	0.3441	0.3800
(191.61°C)	*h*	2808.0	3044.3	3259.7	3475.0	3695.0	3920.8	4152.7
	s	6.5394	6.9938	7.3404	7.6385	7.9060	8.1508	8.3775
14	*v*	0.1429	0.1823	0.2177	0.2520	0.2859	0.3194	0.3528
(195.04°C)	*h*	2801.4	3041.6	3258.2	3473.9	3694.1	3920.2	4152.2
	s	6.4941	6.9561	7.3045	7.6032	7.8710	8.1161	8.3430
15	*v*	0.1324	0.1697	0.2029	0.2359	0.2667	0.2980	0.3292
(198.29°C)	*h*	2794.7	3038.9	3256.6	3472.8	3693.3	3919.6	4151.7
	s	6.4508	6.9207	7.2709	7.5703	7.8385	7.0838	8.3108
16	*v*		0.1587	0.1900	0.2202	0.2499	0.2793	0.3086
(201.37°C)	*h*		3036.2	3255.0	3471.7	3692.5	3918.9	4151.3
	s		6.8873	7.2394	7.5395	7.8080	8.0535	8.2807
17	*v*		0.1489	0.1785	0.2070	0.2351	0.2628	0.2904
(204.31°C)	*h*		3033.5	3253.5	3470.6	3691.7	3918.3	4150.8
	s		6.8557	7.2098	7.5105	7.7793	8.0250	8.2523
18	*v*		0.1402	0.1684	0.1954	0.2219	0.2481	0.2742
(207.11°C)	*h*		3030.7	3251.9	3469.5	3690.9	3917.7	4150.3
	s		6.8257	7.1816	7.4830	7.7522	7.9981	8.2256
19	*v*		0.1325	0.1593	0.1849	0.2101	0.2350	0.2597
(209.80°C)	*h*		3027.9	3250.3	3468.4	3690.0	3917.1	4149.9
	s		6.7970	7.1550	7.4570	7.7265	7.9727	8.2003
20	*v*		0.1255	0.1511	0.1756	0.1995	0.2232	0.2467
(212.37°C)	*h*		3025.0	3248.7	3467.2	3689.2	3916.5	4149.4
	s		6.7696	7.1296	7.4323	7.7022	7.9485	8.1763

Continues

Table A4 Continued

Pressure (bar) (Saturation temperature)		Temperature (°C) 300	400	500	600	700	800
22	v	0.113 43	0.137 00	0.159 34	0.181 19	0.202 76	0.224 15
(217.24°C)	h	3019.3	3245.5	3465.1	3687.6	3915.2	4148.4
	s	6.7179	7.0821	7.3862	7.6568	7.9035	8.1316
24	v	0.103 36	0.125 22	0.145 82	0.165 92	0.185 75	0.205 39
(221.78°C)	h	3013.4	3242.3	3462.9	3685.9	3914.0	4147.5
	s	6.6699	7.0384	7.3439	7.6152	7.8624	8.0908
26	v	0.094 83	0.115 26	0.134 38	0.153 01	0.171 35	0.189 52
(226.04°C)	h	3007.4	3239.0	3460.6	3684.3	3912.7	4146.6
	s	6.6249	6.9979	7.3048	7.5768	7.8245	8.0531
28	v	0.087 51	0.106 71	0.124 58	0.141 94	0.159 02	0.175 91
(230.05°C)	h	3001.3	3235.8	3458.4	3682.6	3911.5	4145.6
	s	6.5824	6.9601	7.2685	7.5412	7.7893	8.0183
30	v	0.081 16	0.099 31	0.116 08	0.132 34	0.148 32	0.164 12
(233.84°C)	h	2995.1	3232.5	3456.2	3681.0	3910.3	4144.7
	s	6.5422	6.9246	7.2345	7.5079	7.7564	7.9857
32	v	0.075 59	0.092 83	0.108 65	0.123 95	0.138 97	0.153 81
(237.45°C)	h	2988.7	3229.2	3454.0	3679.3	3909.0	4143.8
	s	6.5037	6.8912	7.2026	7.4767	7.7257	7.9552
34	v	0.070 68	0.087 11	0.102 09	0.116 57	0.130 71	0.144 70
(240.88°C)	h	2982.2	3225.9	3451.7	3677.7	3907.8	4142.8
	s	6.4669	6.8595	7.1724	7.4473	7.6967	7.9266
36	v	0.066 30	0.082 02	0.096 26	0.109 96	0.123 37	0.136 61
(244.16°C)	h	2975.6	3222.5	3449.5	3676.1	3906.5	4141.9
	s	6.4315	6.8294	7.1439	7.4195	7.6693	7.8995
38	v	0.062 37	0.077 47	0.091 04	0.104 06	0.116 81	0.129 37
(247.31°C)	h	2968.9	3219.1	3447.2	3674.4	3905.3	4141.0
	s	6.3973	6.8007	7.1168	7.3931	7.6434	7.8739
40	v	0.058 83	0.073 38	0.086 34	0.098 76	0.110 90	0.122 85
(250.33°C)	h	2962.0	3215.7	3445.0	3672.8	3904.1	4140.0
	s	6.3642	6.7733	7.0909	7.3680	7.6187	7.8495
45	v	0.051 34	0.064 72	0.076 43	0.087 43	0.098 43	0.109 10
(257.41°C)	h	2944.2	3207.1	3439.3	3668.6	3901.0	4137.7
	s	6.2852	6.7093	7.0311	7.3100	7.5619	7.7934
50	v	0.045 30	0.057 79	0.068 49	0.078 62	0.088 45	0.098 09
(263.91°C)	h	2925.5	3198.3	3433.7	3664.5	3897.9	4135.3
	s	6.2105	6.6508	6.9770	7.2578	7.5108	7.7431

Continues

Table A4 Continued

Pressure (bar) (Saturation temperature)		Temperature (°C) 300	400	500	600	700	800
60	v	0.036 14	0.047 38	0.056 59	0.065 18	0.073 48	0.081 59
(275.55°C)	h	2885.0	3180.1	3422.2	3656.2	3891.7	4130.7
	s	6.0692	6.5462	6.8818	7.1664	7.4217	7.6554
70	v	0.029 46	0.039 92	0.048 09	0.055 59	0.062 79	0.069 80
(285.79°C)	h	2839.4	3161.2	3410.6	3647.9	3885.4	4126.0
	s	5.9327	6.4536	6.7993	7.0880	7.3456	7.5808
80	v	0.024 26	0.034 31	0.041 70	0.048 39	0.054 77	0.060 96
(294.97°C)	h	2786.8	3141.6	3398.8	3639.5	3879.2	4121.3
	s	5.7942	6.3694	6.7262	7.0191	7.2790	7.5158
90	v		0.030 73	0.037 64	0.042 80	0.048 53	0.054 08
(303.31°C)	h		3125.3	3389.2	3631.1	3873.0	4116.7
	s		6.3067	6.6728	6.9574	7.2196	7.4579
100	v		0.026 41	0.032 76	0.038 32	0.043 55	0.048 58
(310.96°C)	h		3099.9	3374.6	3622.7	3866.8	4112.0
	s		6.2182	6.5994	6.9013	7.1660	7.4058
110	v		0.023 51	0.029 50	0.034 66	0.039 47	0.044 08
(318.05°C)	h		3077.8	3362.2	3614.2	3860.5	4107.3
	s		6.1483	6.5432	6.8499	7.1170	7.3584
120	v		0.021 08	0.026 73	0.031 60	0.036 07	0.040 33
(324.65°C)	h		3054.8	3349.6	3605.7	3854.3	4102.7
	s		6.0810	6.4906	6.8022	7.0718	7.3147
130	v		0.019 02	0.024 40	0.029 02	0.033 19	0.037 16
(330.83°C)	h		3030.7	3336.8	3597.1	3848.0	4098.0
	s		6.0155	6.4409	6.7577	7.0298	7.2743
140	v		0.017 23	0.022 51	0.026 90	0.030 72	0.034 44
(336.64°C)	h		3005.6	3323.8	3588.5	3841.7	4093.3
	s		5.9513	6.3937	6.7159	6.9906	7.2367
150	v		0.015 66	0.020 80	0.024 88	0.028 59	0.032 09
(342.13°C)	h		2979.1	3310.6	3579.8	3835.4	4088.6
	s		5.8876	6.3487	6.6764	6.9536	7.2013
175	v		0.012 46	0.017 36	0.021 72	0.025 07	0.028 21
(354.64°C)	h		2906.3	3276.5	3562.2	3822.8	4079.3
	s		5.7274	6.2432	6.6031	6.8857	7.1366
200	v		0.009 947	0.014 77	0.018 16	0.021 11	0.023 85
(365.70°C)	v		2820.5	3241.1	3535.5	3803.8	4065.3
	s		5.5585	6.1456	6.5043	6.7953	7.0511

Continues

Table A4 Continued

Pressure (bar) (Saturation temperature)		Temperature (°C) 400	500	600	700	800
220	v	0.008 251	0.013 12	0.016 33	0.019 07	0.021 60
(373.69°C)	h	2738.8	3211.7	3517.4	3791.1	4055.9
	s	5.4102	6.0716	6.4441	6.7410	7.0001
250	v	0.006 014	0.011 13	0.014 13	0.016 63	0.018 91
	h	2582.0	3165.9	3489.9	3771.9	4041.9
	s	5.1455	5.9655	6.3604	6.6664	6.9306
300	v	0.002 831	0.008 681	0.011 44	0.013 65	0.015 62
	h	2161.8	3085.0	3443.0	3739.7	4018.5
	s	4.4896	5.7975	6.2340	6.5560	6.8288
400	v	0.001 909	0.005 616	0.008 088	0.009 930	0.011 52
	h	1934.1	2906.8	3346.4	3674.8	3971.7
	s	4.1190	5.4762	6.0135	6.3701	6.6606
500	v	0.001 729	0.003 882	0.006 111	0.007 720	0.009 076
	h	1877.7	2723.0	3248.3	3610.2	3925.3
	s	4.0083	5.1782	5.8207	6.2138	6.5222
1000	v	0.001 446	0.001 893	0.002 668	0.003 536	0.004 341
	h	1797.6	2316.1	2857.5	3324.4	3714.3
	s	3.7738	4.4913	5.1505	5.6579	6.0397

Table A5 Saturated Freon-12: Temperature table

T (°C)	P^s kPa	v_f (m³/kg)	v_g (m³/kg)	h_f (kJ/kg)	h_g (kJ/kg)	s_f (kJ/kg K)	s_g (kJ/kg K)
−40.0	64.123	0.000 659	0.242 233	0.0000	169.7004	0.0000	0.7279
−38.0	70.376	0.000 662	0.222 125	1.7759	170.6318	0.0076	0.7256
−36.0	77.100	0.000 664	0.204 016	3.5560	171.5613	0.0151	0.7235
−34.0	84.319	0.000 667	0.187 683	5.3402	172.4887	0.0226	0.7215
−32.0	92.057	0.000 669	0.172 920	7.1287	173.4137	0.0300	0.7195
−30.0	100.341	0.000 672	0.159 553	8.9215	174.3361	0.0374	0.7177
−28.0	109.193	0.000 675	0.147 433	10.7185	175.2557	0.0447	0.7159
−26.0	188.643	0.000 677	0.136 420	12.5201	176.1723	0.0520	0.7142
−24.0	128.715	0.000 680	0.126 400	14.3261	177.0856	0.0592	0.7125
−22.0	139.436	0.000 683	0.117 268	16.1366	177.9955	0.0665	0.7109
−20.0	150.836	0.000 685	0.108 931	17.9517	178.9017	0.0736	0.7094
−18.0	162.940	0.000 688	0.101 311	19.7713	179.8040	0.0807	0.7080
−16.0	175.778	0.000 691	0.094 335	21.5957	180.7022	0.0878	0.7066
−14.0	189.379	0.000 694	0.087 938	23.4249	181.5959	0.0949	0.7052
−12.0	203.771	0.000 697	0.082 065	25.2589	182.4851	0.1019	0.7039
−10.0	218.985	0.000 700	0.076 667	27.0978	183.3695	0.1089	0.7027
−8.0	235.050	0.000 703	0.071 697	28.9417	184.2487	0.1158	0.7015
−6.0	251.998	0.000 706	0.067 115	30.7908	185.1227	0.1227	0.7004
−4.0	269.859	0.000 709	0.062 887	32.6450	185.9911	0.1296	0.6993
−2.0	288.661	0.000 712	0.058 981	34.5045	186.8538	0.1364	0.6983
0.0	308.440	0.000 716	0.055 366	36.3695	187.7104	0.1432	0.6973
2.0	329.222	0.000 719	0.052 019	38.2398	188.5608	0.1500	0.6963
4.0	351.047	0.000 723	0.048 914	40.1161	189.4095	0.1567	0.6954
6.0	373.944	0.000 726	0.046 033	41.9981	190.2414	0.1634	0.6945
8.0	397.942	0.000 730	0.043 355	43.8861	191.0712	0.1701	0.6936
10.0	423.083	0.000 733	0.040 863	45.7805	191.8935	0.1768	0.6928
12.0	449.392	0.000 737	0.038 542	47.6811	192.7082	0.1834	0.6920
14.0	476.905	0.000 741	0.036 379	49.5883	193.5149	0.1900	0.6912
16.0	505.660	0.000 744	0.034 361	51.5024	194.3132	0.1965	0.6904
18.0	535.684	0.000 748	0.032 476	53.4234	195.1030	0.2031	0.6897

Continues

Table A5 Continued

T (°C)	P^s kPa	v_f (m^3/kg)	v_g (m^3/kg)	h_f (kJ/kg)	h_g (kJ/kg)	s_f (kJ/kg K)	s_g (kJ/kg K)
20.0	567.019	0.000 752	0.030 713	55.3518	195.8836	0.2096	0.6890
22.0	599.698	0.000 756	0.029 063	57.2878	196.6550	0.2161	0.6883
24.0	633.756	0.000 761	0.027 518	59.2318	197.4165	0.2226	0.6876
26.0	669.233	0.000 765	0.026 069	61.1841	198.1679	0.2290	0.6869
28.0	706.156	0.000 769	0.024 709	63.1449	198.9088	0.2355	0.6863
30.0	744.567	0.000 774	0.023 432	65.1147	199.6387	0.2419	0.6856
32.0	784.508	0.000 778	0.022 232	67.0939	200.3571	0.2483	0.6850
34.0	826.003	0.000 783	0.021 103	69.0827	201.0637	0.2547	0.6844
36.0	869.108	0.000 788	0.020 040	71.0821	201.7577	0.2611	0.6838
38.0	913.842	0.000 793	0.019 038	73.0921	202.4388	0.2674	0.6831
40.0	960.255	0.000 798	0.018 093	75.1134	203.1063	0.2738	0.6825
42.0	1008.379	0.000 803	0.017 202	77.1464	203.7598	0.2801	0.6819
44.0	1058.266	0.000 808	0.016 359	79.1921	204.3982	0.2865	0.6813
46.0	1109.948	0.000 814	0.015 563	81.2510	205.0212	0.2928	0.6806

Table A6 Saturated Freon-12: Pressure table

P^s (kPa)	T (°C)	v_f (m³/kg)	v_g (m³/kg)	h_f (kJ/kg)	h_g (kJ/kg)	s_f (kJ/kg K)	s_g (kJ/kg K)
60	− 41.4	0.000 658	0.257 692	− 1.2435	169.0452	− 0.0053	0.7295
65	−39.7	0.000 660	0.239 187	0.2570	169.8353	0.0011	0.7275
70	−38.1	0.000 662	0.223 235	1.6738	170.5776	0.0071	0.7258
75	−36.6	0.000 664	0.209 336	3.0126	171.2783	0.0128	0.7242
80	−35.2	0.000 665	0.197 114	4.2870	171.9419	0.0182	0.7227
85	−33.8	0.000 667	0.186 280	5.5033	172.5727	0.0232	0.7213
90	−32.5	0.000 669	0.176 606	6.6651	173.1741	0.0281	0.7200
95	−31.3	0.000 670	0.167 915	7.7781	173.7488	0.0327	0.7188
100	−30.1	0.000 672	0.160 061	8.8499	174.2994	0.0371	0.7177
110	−27.8	0.000 675	0.146 421	10.8768	175.3365	0.0453	0.7157
120	−25.7	0.000 678	0.134 975	12.7702	176.2990	0.0530	0.7139
130	−23.8	0.000 680	0.125 229	14.5484	177.1980	0.0601	0.7123
140	−21.9	0.000 683	0.116 825	16.2280	178.0418	0.0668	0.7108
150	−20.1	0.000 685	0.109 501	17.8234	178.8372	0.0731	0.7095
160	−18.5	0.000 688	0.103 060	19.3387	179.5901	0.0791	0.7083
170	−16.9	0.000 690	0.097 349	20.7888	180.3049	0.0847	0.7072
180	−15.4	0.000 692	0.092 249	22.1743	180.9855	0.0901	0.7061
190	−13.9	0.000 694	0.087 667	23.5064	181.6354	0.0952	0.7052
200	−12.5	0.000 696	0.083 526	24.7868	182.2572	0.1001	0.7043
215	−10.5	0.000 699	0.078 010	26.6246	183.1430	0.1071	0.7030
230	−8.6	0.000 702	0.073 187	28.3730	183.9779	0.1137	0.7019
245	−6.8	0.000 705	0.068 933	30.0393	184.7681	0.1199	0.7009
260	−5.1	0.000 708	0.065 152	31.6329	185.5182	0.1258	0.6999
275	−3.4	0.000 710	0.061 768	33.1636	186.2321	0.1315	0.6990
290	−1.9	0.000 713	0.058 721	34.6338	186.9134	0.1369	0.6982
305	−0.3	0.000 715	0.055 962	36.0525	187.5649	0.1421	0.6974
320	−1.1	0.000 718	0.053 453	37.4205	188.1895	0.1470	0.6967
340	3.0	0.000 721	0.050 438	39.1775	188.9833	0.1534	0.6958
360	4.8	0.000 724	0.037 746	40.8629	189.7373	0.1594	0.6950
380	6.5	0.000 727	0.045 326	42.4823	190.4552	0.1651	0.6943
400	8.2	0.000 730	0.043 139	44.0433	191.1402	0.1707	0.6935
420	9.8	0.000 733	0.041 153	45.5527	191.7952	0.1760	0.6929
440	11.3	0.000	0.039 340	47.0118	192.4225	0.1810	0.6923
460	12.8	0.000 738	0.037 679	48.4253	193.0246	0.1859	0.6917
480	14.2	0.000 741	0.036 151	49.7983	193.6028	0.1907	0.6911

Continues

Table A6 Continued

P^s (kPa)	T (°C)	v_f (m³/kg)	v_g (m³/kg)	h_f (kJ/kg)	h_g (kJ/kg)	s_f (kJ/kg K)	s_g (kJ/kg K)
500	15.6	0.000 744	0.034 741	51.1321	194.1595	0.1953	0.6906
530	17.6	0.000 748	0.032 817	53.0672	194.9566	0.2019	0.6898
560	19.6	0.000 751	0.031 091	54.9270	195.7124	0.2082	0.6891
590	21.4	0.000 755	0.029 534	56.7215	196.4303	0.2142	0.6885
620	23.2	0.000	0.028 122	58.4559	197.1138	0.2200	0.6879
650	24.9	0.000 763	0.026 835	60.1346	197.7655	0.2256	0.6873
680	26.6	0.000 766	0.025 658	61.7636	198.3883	0.2309	0.6867
710	18.2	0.000 770	0.024 575	63.3452	198.9836	0.2361	0.6862
740	29.8	0.000 773	0.023 578	64.8837	199.5539	0.2411	0.6857
770	31.3	0.000 777	0.022 654	66.3828	200.1010	0.2460	0.6852
800	32.8	0.000 780	0.021 797	67.8449	200.6255	0.2507	0.6848
840	34.7	0.000 785	0.020 746	69.7396	201.2931	0.2568	0.6842
880	36.5	0.000 789	0.019 787	71.5775	201.9270	0.2626	0.6836
920	38.3	0.000 793	0.018 907	73.3639	202.5294	0.2683	0.6831
960	40.0	0.000 798	0.018 098	75.1013	203 1030	0.2737	0.6825
1000	41.7	0.000 802	0.017 351	76.7967	203.6486	0.2790	0.6820
1050	43.7	0.000 808	0.016 494	78.8585	204.2498	0.2854	0.6814
1100	45.6	0.000 813	0.015 711	80.8591	204.9043	0.2916	0.6808
1200	49.3	0.000 824	0.014 333	84.7076	206.0212	0.3034	0.6792
1300	52.8	0.000 834	0.013 157	88.3725	207.0166	0.3144	0.6784

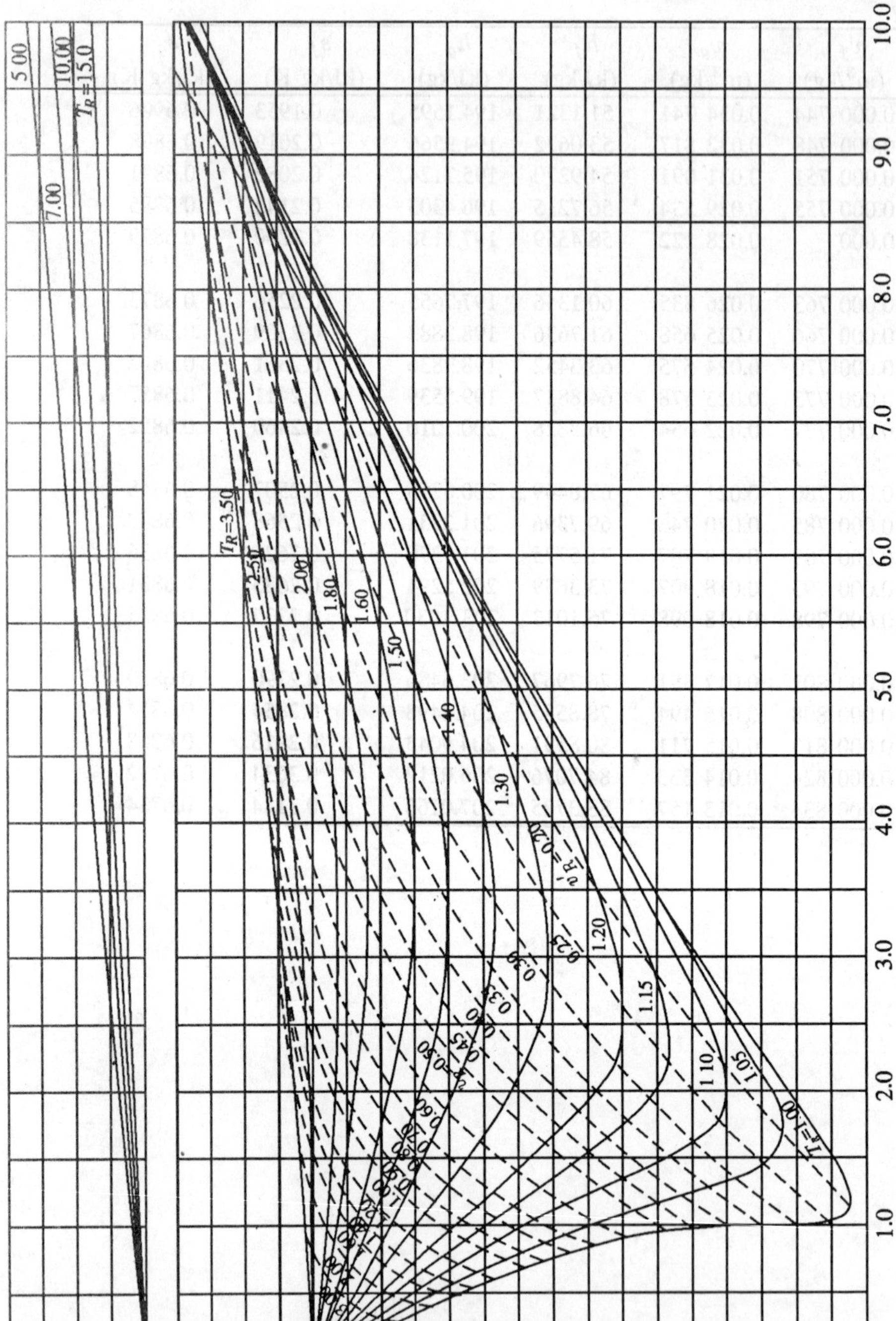

Fig. A1 Compressibility chart, low pressure range.

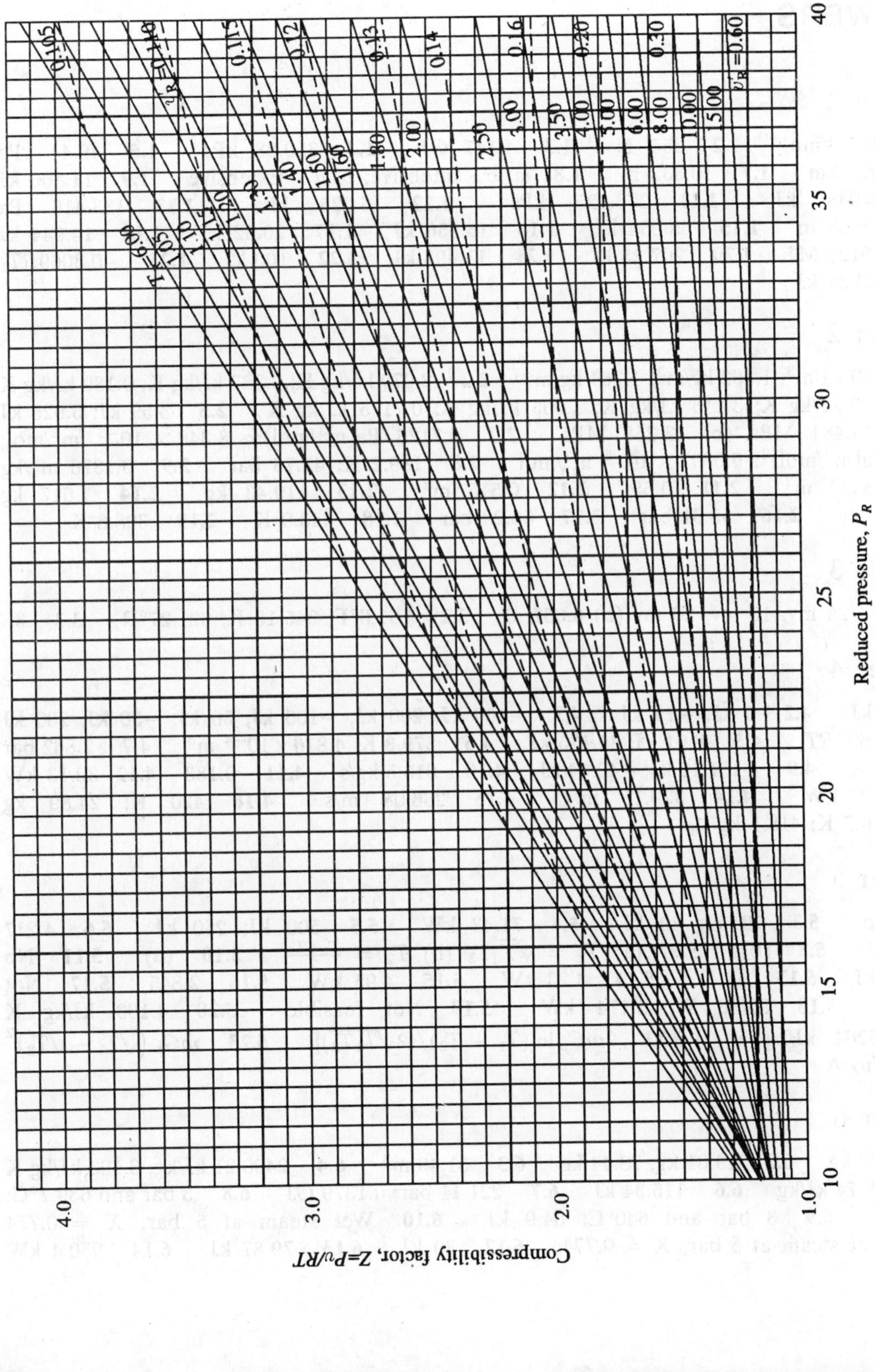

***Fig. A2** Compressibility chart, high pressure range.*

ANSWERS

Chapter 1

1.1 3.467 kmol **1.2** 255 N **1.3** 4.147 GJ **1.4** 350.88 kPa **1.5** 49.44 kPa **1.6** 8.904 km **1.7** 10.332 m **1.8** *a, c* - intensive; *b, d* - Extensive **1.9** 11.806 kJ **1.10** 53.048 kJ **1.11** 276.325 kPa **1.12** 16.325 kPa **1.13** 133.416 Pa **1.14** 509.68 m **1.15** 53.054 kJ **1.16** 16.856 kJ **1.17** 20.52 kJ **1.18** 14.944 kJ **1.19** 3.9128 MJ **1.20** 207.25 kJ **1.21** 11.493 kJ **1.22** 100 kJ **1.23** $-0.3069RT_1$ **1.24** 163.28 kJ

Chapter 2

2.1 0.160 kg/m^3; 1.162 kg/m^3; 1.764 kg/m^3 **2.2** 2.079 kJ/kg K; 0.287 kJ/kg K; 0.189 kJ/kg K **2.3** 5.195 kJ/kg K; 3.116 kJ/kg K; 1.005 kJ/kg K; 0.7175 kJ/kg K **2.5** 3.59 kJ; 5.025 kJ **2.6** (a)15.484 MPa (b) 12.258 MPa **2.7** 0.7126 Pa(m^3/mol)^2; 8.348×10^{-5} m^3/mol; 0.9373 Pa(m^3/mol)^2; 9.0303×10^{-5} m^3/mol **2.8** 189.2 K; 45.73 bar **2.9** 0.1376 m^3/kg **2.10** 0.8436 m^3 **2.11** 0.55 **2.12** 0.515 m^3 **2.13** 119.43 kg **2.14** 1.617 kg **2.15** 20 kg **2.16** 58.61 bar **2.17** 56.29 bar **2.18** 443.9 K **2.19** 308.6 K

Chapter 3

3.1 (a) −25 mv; 15 mv; 20 mv (b) 6.667; 0 **3.2** 161.6°F; 345.15 K; 621.27°R **3.3** 3

Chapter 4

4.1 25 kJ **4.2** 0 kJ; −15 kJ **4.3** −100 kJ; 200 kJ; −100 kJ; 50 kJ; −50 kJ; 200 kJ **4.4** $-1.693RT_1$ **4.5** 8.61 kJ; 30.135 kJ **4.6** 579.3 K; 4.816×10^{-4} m^3 **4.7** 2.862 bar **4.8** 246 K **4.9** −101.12 kJ; −40.64 kJ **4.10** 616.5 kg/s **4.11** 0.183 **4.12** 30.13 kW **4.13** 580 kW **4.14** 329.15 m/s **4.15** 266.08 m/s **4.16** 420 K; 24.89 kg **4.17** 416.7 K; 418.1 kg

Chapter 5

5.1 No **5.3** 70 kJ; 30 kJ **5.4** 71.43 kW **5.5** 500 kJ; 250 kJ **5.6** 4.217 **5.7** 14.9 **5.8** (b) **5.9** (a) $T_2 = \sqrt{T_1 T_3}$ (b) $T_2 = \frac{T_1+T_3}{2}$ **5.10** (a) **5.11** No **5.12** 3 kJ **5.13** 45 W **5.14** 4.23 kW **5.15** 4.95 kW **5.16** 2.845 **5.17** Not justified **5.18** 966.6 W; 2.174 kW **5.19** Not feasible **5.20** 4.103 kJ/kg K **5.21** 7.5264 kJ/kg K **5.22** $2mC_p \ln[(T_1 + T_2)/(2\sqrt{T_1 T_2})]$ **5.23** $mC_p\left(\sqrt{T_1} - \sqrt{T_2}\right)^2$ **5.24** B to A

Chapter 6

6.1 78.62 kg **6.2** 19.61 kg; 78.44 kg **6.3** 51.49 m^3 **6.4** 2400.21 kJ/kg; 6.193 kJ/kg K **6.5** 2245.74 kJ/kg **6.6** 115.34 kJ **6.7** 221.11 bar; 0.1379 GJ **6.8** 3 bar and 639.7°C; 55.07 kJ **6.9** 3 bar and 640°C; 54.9 kJ **6.10** Wet steam at 5 bar; $X = 0.774$ **6.11** Wet steam at 5 bar; X = 0.774 **6.12** 80 kJ **6.13** 79.87 kJ **6.14** 950.4 kW

6.15 950 kW **6.16** 377.8 m/s **6.17** 376.8 m/s **6.18** 0.976 **6.19** 0.976 **6.20** 0.935 **6.21** 0.935 **6.22** 0.612 **6.23** 0.613 **6.24** 0.362; 276 kg/s **6.25** 0.2894; 345.4 kg/s **6.26** 0.2677; 256 kg/s

Chapter 7

7.1 (a) 689.2 K; 18.38 bar; 3476.7 K; 92.72 bar; 1513.3 K; 5.04 bar; (b) 0.565 (c) 1130 kJ **7.2** 1.5 MPa **7.3** 0.565; 472.96 kJ/kg air **7.4** 6.35 **7.5** 0.536 **7.6** 1.269 MPa **7.7** $(T_3/T_1)^{1/2(\gamma-1)}$ **7.8** 0.604 **7.9** 8.2 bar **7.10** 13.942; 2.387; 0.427 **7.12** 15.59; 1.995; 1795.5 K **7.13** 948.3 K; 180 kJ **7.14** 11.313; 0.5 **7.15** 8.223 **7.16** 317.37 kJ/kg air; 728.48 kJ/kg air

Chapter 8

8.1 Increase the evaporator temperature **8.2** $[mc(T_1 - T_2)^2]/T_2$ **8.3** 4.669 kW **8.4** 2.54; 2.77 kW **8.5** 2.599; 13.02 tons; 17.62 kW

Chapter 9

9.1 (a) 2.5446 kPa; (ab) 174.54 kg (c) 10904.8 (d) 21.3°C **9.2** 34°C **9.3** 0.0203 kg water vapor/kg dry air; 0.434 **9.4** 29.8°C **9.5** 0.0158 kg water vapor/kg dry air; 0.446 **9.6** 0.0176 kg water vapor/kg dry air; 0.496 **9.7** 0.0178 kg water vapor/kg dry air; 0.505 **9.8** 0.0235 kg water vapor/kg dry air; 0.5; 101 kJ/kg dry air **9.9** 34°C; 0.0201 kg water vapor/kg dry air **9.10** 0.18; 1.75 kg dry air/s **9.11** 17°C; 0.76 **9.12** 0.018 kg water vapor/kg dry air; 0.6 **9.13** 0.0075 kg water vapor/kg dry air; 0.157 kg **9.14** 548.26 kW; 0.14 kg/s **9.15** 0.115 kg dry air/s **9.16** 703.5 kW; 150 kW **9.17** 25°C; 172.8 kg **9.18** 24.5°C; 1404 kg **9.19** 31°C **9.20** 60.09°C; 12.396 kPa

INDEX